Management for Professionals

More information about this series at http://www.springer.com/series/10101

Xiaoming Zhu

China's Technology Innovators

Selected Cases on Creating and Staying
Ahead of Business Trends

Xiaoming Zhu
China Europe International Business School
Shanghai
China

ISSN 2192-8096   ISSN 2192-810X   (electronic)
Management for Professionals
ISBN 978-981-10-5387-0   ISBN 978-981-10-5388-7   (eBook)
DOI 10.1007/978-981-10-5388-7

Jointly published with Shanghai Jiao Tong University Press

The print edition is not for sale in China Mainland. Customers from China Mainland please order the print book from: Shanghai Jiao Tong University Press.

Library of Congress Control Number: 2017944541

Printed on acid-free paper

This Springer imprint is published by Springer Nature
The registered company is Springer Nature Singapore Pte Ltd.
The registered company address is: 152 Beach Road, #21-01/04 Gateway East, Singapore 189721, Singapore

Foreword I

The Inevitable Trend of an Innovation-driven Future

Professor Xiaoming Zhu always says, "Innovation will drive the trends of the future." Innovation has now become a key platform for national growth, as well as a powerful engine for social and industrial development. A range of new technologies, products, and business models have emerged as we enter the era of the "Internet of Everything". While this offers many inspiring opportunities, you can also be overwhelmed on the pathway to innovation. To succeed in an ever-changing world, and make progress with specific goals, we need insight into the future and a sharp mind for business and technological development.

Professor Xiaoming Zhu has always been committed to exploring the frontiers of innovation, despite the demands of his administrative activities. He is known as the "24/7 President and Professor" of CEIBS as he is always working, even after hours, on weekends and during holidays. His lectures are rich in content and passionately delivered, making him the most popular professor at CEIBS. After several works and translations such as *Payment Revolution*, *Mastering the Hype Cycle*, and *Business Trends in the Digital Age: Evolution of Theories and Applications*, Prof. Xiaoming Zhu has completed *China's Technology Innovators: Selected Cases on Creating and Staying Ahead of Business Trends*, which presents various successful innovation case studies about Chinese companies, such as *3DMed: R&D Platform for Personalized Anti-cancer Precision Drugs*, *Alibaba: The Decade-long Road to Financial Services*, *Flexible Adaptation: "Autonomous Navigation" of Amap*, and *ICBC in the Digital Times*, all representative case studies that Prof. Xiaoming Zhu has used for teaching in the last 2–3 years, and continuously updated and improved. All the case studies have a shared focus: how to achieve successful entrepreneurship and innovation in the primary, secondary, and tertiary sectors through Big Data, cloud computing, the platform economy, and mobile Internet. As well as a detailed description of the case, all of the case studies include interpretations and comments from experts in relevant areas and EMBA students, which are particularly helpful to readers, enabling in-depth study of the cases from multiple perspectives.

I am pleased to see that CEIBS, as the leading Chinese business school in management education for the past 20 years, is still committed to the principles of exploration and innovation, with a focus on "China Depth and Global Breadth", thanks to the tireless efforts of all the management, faculty, and students. CEIBS will work together with Chinese companies during their future journey of innovation.

October 2015

Dr. Li Mingjun
President and Professor of Management
China Europe International Business School
Shanghai, China

Foreword II

This book of cases on innovation practices of Chinese companies in the digital era is another important step forward in the process of building a strong base of knowledge on the research of Chinese innovation in China.

Teaching by the case method is complex. Peter Drucker, the leading philosopher in the field of management, proposed that management professors should develop frameworks to address professionally the management issues of their interest. Cases become the vehicle to bring the real world to the classroom and practice the application of those frameworks. This is a method that can help a lot to develop new management applicable knowledge.

CEIBS pioneered in China a relevant effort in case writing as well as in the use of cases in all the portfolio of programs the school offers around the world. We can argue that management is a global discipline, and that what happens inside a classroom in America, in Europe, or in China, must, therefore, be very similar. This means that business schools around the world should discuss cases on issues that happen in Europe, in America, in China, or in other relevant areas of the world.

The objective of this book is to contribute to fill the gap between the interest of the world's management community in China and the limited availability of cases on innovation issues in China. Moreover, in the background of the fast development of mobile Internet, China sets off a wave of popular entrepreneurship and innovation and many companies would face the difficulties of transformation and business model changing. The publication of this book is a good reference for their problems' resolving.

These cases would have not been possible without the generous collaboration of the management of the different companies involved. They dedicated time to and shared information with the case writers. By doing this, these companies share with the case writers and the publishers of the book the merits of this most needed contribution. Meanwhile, we will also thank CEIBS EMBA alumni and professors from CEIBS whose dedicated work has narrowed the distance between the case authors and the reader.

September 2015

Prof. Pedro Nueno
Executive President
China Europe International Business School
Shanghai, China

Author's Preface

At the turn of 2015, *Mastering the Hype Cycle* saw its Chinese translation issued; the monography *Business Trends in the Digital Age: Evolution of Theories and Applications* (Chinese version) was published. Based on these books, my 2-day lecture, "Business Trends and Technology Innovation", commands great popularity with CEIBS MBA and FMBA (Finance MBA) students, who often lend me a helping hand with case writing. Over the past 5 years, I have devoted much time to topic selection, company visit, and case writing, while authoring books and preparing teaching materials as a full-time professor.

My new book, *China's Technology Innovators: Selected Cases on Creating and Staying Ahead of Business Trends* (Chinese version), is due for publication at the turn of 2016. When correcting the final proofs one month ago, I wrote this preface to provide readers with some enlightenment: (1) "Sorting the Wheat from the Chaff" indicates there is strict underlying logic behind ten mega business trends detailed in this book; (2) "A Straw in the Wind" implies keeping abreast of technology innovation helps us identify business trends; (3) "As Compelling as Ever" is an ultimate target for classroom lectures, academic studies or case teaching.

Sorting the Wheat from the Chaff

In the times of digital technology and Internet, ideas about user orientation, iteration, and platform, freemium and sharing have caught on; the tide of openness and inclusiveness, deconstructing, and disintermediation runs strong; garage culture, café culture, and grassroots culture have taken hold all over the world.

Which underlying business trends merit special attention? In my mind, ten mega business trends in the digital age described in *Business Trends in the Digital Age* demonstrate sound logic: Trend I, Trend II, Trend III, and Trend IV, "Big Data, Cloud Computing, Platforms and Mobile Internet", predict the changes in infrastructure; Trend V, "Software-Defined Anything", focuses on how software will become the most important force of production in the digital age; Trend VI, namely,

"Outsourcing and Crowdsourcing", anticipates changes in production organization; Trend VII, "Driven By Supply and Demand", explains the changing forces of economic growth; Trend VIII, "Long Tail", forecasts an altered competitive landscape; Trend IX, "Digital Finance", analyzes changes in financial entities and models; Trend X, "Collaboration", probes into changes in operations models.

In this book, Case V: Alibaba: A Decade-Long Road to Financial Services and Case VI: Changing with the Times: AutoNavi's Autonomous Development will give you some enlightenment as to how to gain a solid foothold in the market amidst fierce competition.

In the times of digital technology and Internet, it has been the norm that "following trends" outweighs "building up advantages". Thus, those who follow the beaten track breathe a sigh of regret that they have been "out" before getting down to SWOT analysis.

A Straw in the Wind

The path to new business models has never been wider. Technological forecast plays an instrumental role. Every year, the world-renowned consulting services provider Gartner releases the Hype Cycle, which forecasts the trends for cutting-edge technologies.

The entrepreneurs mentioned in this book have realized any company, which gets a head start in technology innovation and is committed to innovation, will gain a sustainable competitive edge. You can draw inspiration from Case I: 3DMed: An R&D Platform for Personalised Precision Anti-cancer Drugs, Case II: Micro Platform, Major Innovation—WeChat-Based Ecosystem of Innovation, Case III: SHANGHAI GM: The Way to Intelligent Manufacturing, Case IV: Can Robots Raise Laying Hens?, and Case VII: ICBC in the Digital Times.

Our case teaching has evolved from writing cases to developing a case database. When delivering the lecture "Business Trends and Technology Innovation", I have integrated the technology's five Stages mentioned in the Hype Cycle, namely, the Technology Trigger, Peak of Inflated Expectations, Trough of Disillusionment, Slope of Enlightenment, and Plateau of Productivity, with the IPO's five rounds, namely, the angel round, Series A, Series B/C/D, Pre-IPO, and IPO to compile the "5×5 Table". Many students believe: (1) Their startup process echoes one or several sections in the Table; (2) Training on the "5×5 Table" needs to be provided for entrepreneurs at the very beginning, as they would likely trail behind from the start, without deep insights into technology innovation for project assessment; (3) The Table offers a valuable guide to "investment for startup and innovation". Some industrial associations, consultancies, and investment banks are eager for tie-ups with CEIBS to beef up the case database as a reward for investors and entrepreneurs.

As Compelling as Ever

Over the past 2 years, we have set about supplementing written cases with audio and video materials, which necessitates a good mastery of new media technology. In this sense, professors need to play a role as movie director. When delivering a lecture to FMBA students, we have created the "6M Diamond Model for Competence" with 3ds Max for case teaching, which has produced the desired effect. It is suggested professors develop visualized cases with such software as iMovie, Final Cut, Premiere, Flash, and Screen Flow.

It is my great honor that the Chinese version of *Mastering the Hype Cycle* has been selected into Jiefang Daily's "Ten Best Books of 2015" and *Business Trends in the Digital Age* has received the first prize for academic works at the 4th Chinese College Press Book Awards. We are convinced that case teaching will give lectures and research a new lease of life to make them as compelling as ever.

On this occasion, I would like to express my thanks to case research writers from CEIBS Case Development Center for their continuous support and to CEIBS professors and alumni for their insightful comments, and extend my best regards to President Li Mingjun and President Pedro Nueno for their Forwards.

In the times of digital technology and Internet, everyone has an opportunity to realize their dream. Whether it is the commitment for entrepreneurs and educators and for characters in the cases and case writers, or to follow the ever-changing business trends and keep pace with the fast-growing technology innovation, forge ahead with determination matters. Only by getting ahead of the leader and going beyond ourselves can we make a foray into uncharted territory, greater than the known area, to create a brave new world.

Shanghai, China
October 2015

Dr. Xiaoming Zhu
Professor of Management and Former President
(June 2006–March 2015)
China Europe International Business School

Contents

Case I: 3DMed: An R&D Platform for Personalised Precision Anti-cancer Drugs

In 2010, Xiong Lei, a Ph.D. graduate from the Institute of Biochemistry and Cell Biology, SIBCB, CAS, left his post-doctoral research in the University of Zurich to found 3DMed with five other senior professionals, aiming to substantially improve R&D efficiency and approval rates of anti-cancer drugs and deliver highly effective clinical treatment for cancer. The new company is headquartered in the Shanghai Caohejing Hi-Tech Park, where over 140 biomedical start-ups are planning their business roadmaps. The question now is how 3DMed, which had completed two rounds of financing by May 2014, will achieve the expected performance and leverage talent, marketing and capital management, as well as strategic planning, to address the challenges it faces in the early stage of business development as well as to minimise innovation risks in an emerging industry.

Overview of Anti-cancer Drug R&D

Malignant tumours, generally referred to as cancer, represent a major disease that can cause great harm to human health. According to global statistics published in 2011, cancer causes nearly 13% of total deaths around the world each year.[1] In 2010, there were 3.093 million new cancer cases and 1.956 million cancer-related deaths in China, according to the *Chinese Cancer Registry Annual Report*. With an average life expectancy of 74 years, Chinese people have a 22% cumulative risk of

This case study was co-authored by Professor Zhu Xiaoming, Part-Time Research Assistant Song Yanbo and Research Associate Ni Yingzi of the China-Europe International Business School on the basis of research materials. It was written with the support of 3D Medicines Corporation. The case study is intended to stimulate classroom discussion, and not to analyse how effectively the company is managed.

[1]Jemal et al. [1].

© Springer Nature Singapore Pte Ltd and Shanghai Jiao Tong University Press 2018
X. Zhu, *China's Technology Innovators*, Management for Professionals,
DOI 10.1007/978-981-10-5388-7_1

developing cancer.[2] It is therefore imperative that we place great importance on cancer prevention and treatment in modern society.

With the number of new cancer cases increasing, the global cancer treatment market in 2010 was valued at USD 59.7 billion, accounting for 10% of the total value of the global prescription drug market.[3] Yet it is still an undeniable fact that cancer remains extremely difficult to cure completely. The risk of cancer metastasis and relapse after surgically removing tumours is high, and traditional anti-cancer drugs only have an average success rate of around 20%.[4] In another perspective, 95.3% of new anti-cancer drugs developed by pharmaceutical companies cannot get approval for launch from the Food and Drug Administration (FDA) due to their failure to meet efficacy requirements,[5] resulting in an annual loss of several billion dollars to drug R&D.[6]

History of Anti-cancer Drug R&D

Though earlier attempts to develop anti-cancer drugs are recorded in global literature, it is generally believed that systematic research on anti-cancer drugs started in the 1940s (see Fig. 1). Since then, researchers have been exploring anti-cancer drugs with experiment and projects scattered around the world. While large-scale cancer research was not possible until the 1950s when several research institutes, such as the National Cancer Institute (NCI) and the European Organisation for Research and Treatment of Cancer (EORTC), were established.[7] With the development of molecular oncology, researchers identified uncontrolled cell cycles as the major cause of cancer. The anti-cancer drugs used for clinical treatment in the 1970s and 1980s were developed to inhibit the proliferation of cancer cells, mainly by interfering with cell division in a nonspecific way. Such traditional cytotoxic drugs, however, kill more than just cancer cells. They also destroy quickly dividing normal cells in the human body. Even though they can extend life for several months or years, many patients suffer strong adverse reactions to the drugs.

Then, with the rapid development of molecular and cell biology, researchers became aware of how cancer cells grow and proliferate, as well as their cycle control mechanisms, and discovered that cancer is caused by abnormal cell proliferation resulting from genetic mutations. Genes, formerly known as Mendelian

---

[2]Chen [2].
[3]American Cancer Society [3].
[4]DeSantis et al. [4].
[5]Berkrot [5].
[6]Pammolli et al. [6].
[7]Mao and Chen [7].

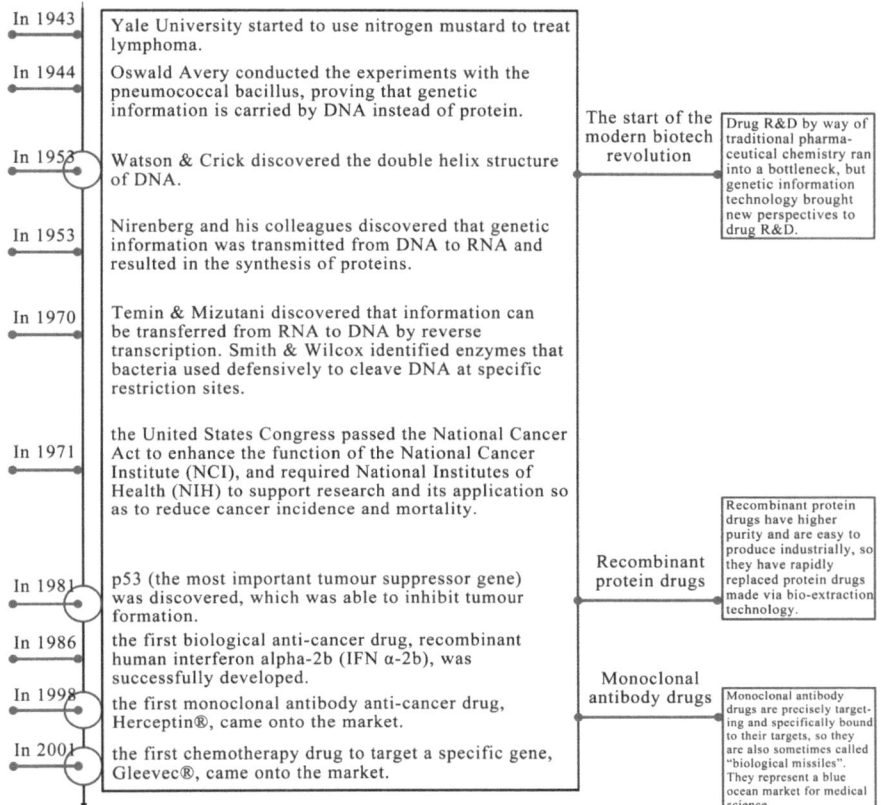

Fig. 1 The history of anti-cancer drug R&D. *Source* Shanghai Center of Novelty Checking and Inquiry, Chinese Academy of Sciences

factors, are DNA sequences that carry genetic information. Genetic information is expressed via the synthesis of proteins, which controls the various characters of an organism. Human cells normally grow and die in a controlled manner. However, when gene mutations are triggered by carcinogens such as cancer-causing chemical substances, radiation or viruses, cell growth in part of a tissue can get out of control, resulting in abnormal proliferation and ultimately cancer.

On this basis, researchers gradually discovered several key enzymes that were related to the differentiation, proliferation and death of cancer cells. These enzymes work on signal transduction pathways in cancer cells and can be targeted by drugs to inhibit the growth of cancer cells in a specific way and reduce negative effects on normal cells, as well as adverse drug reactions. Therefore, since the 1990s, researchers have shifted their focus from cytotoxic drugs and broad-spectrum

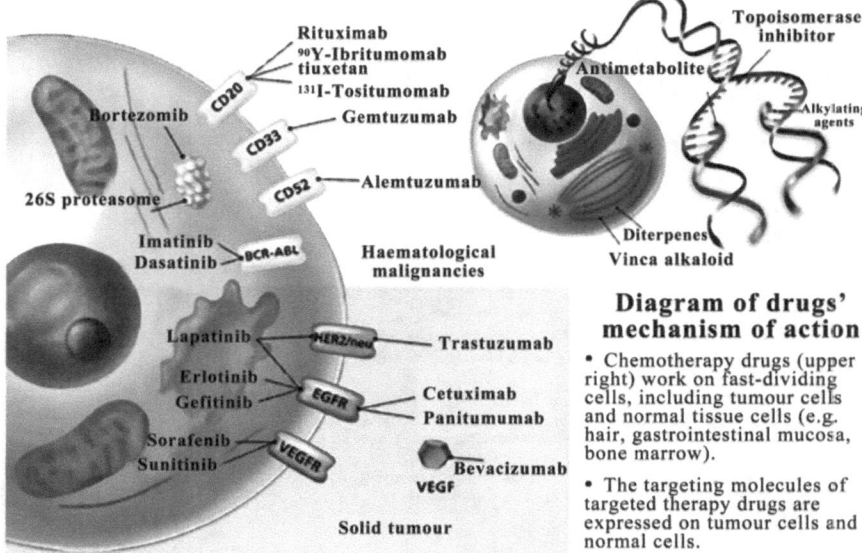

Fig. 2 Mechanism of action of chemotherapy drugs and targeted therapy drugs. *Source* Sun Yan: *The Long Development of Targeted Therapy Drugs: Handling Rationally the Negative Responses of Targeting Drugs*, the website of the Chinese Anti-Cancer Association, http://www.caca.org.cn/system/2009/05/07/010023580.shtml, last view date: 3 November 2015

inhibitors of the cell cycle and DNA metabolism to more specific inhibitors of signal transduction proteins,[8] also known as "targeted therapy drugs" (see Fig. 2). Targeted therapy drugs interfere with specific factors (mainly proto-oncogenes) needed for oncogenesis and reverse the malignant phenotype of cancer cells, thus inhibiting the growth of cancer cells and even eliminating them. They can be divided into two categories, namely "small molecule compounds" and "monoclonal antibodies", according to different molecular weight.

From R&D to Launch

Under the traditional model of drug R&D around the turn of the millennium, new drugs need to undergo four major stages from primary lab studies to commercial sale in pharmacies (see Fig. 3).

Lab studies are the first stage. Pharmaceutical companies must choose a disease condition and a mechanism of action for the new drug. After drug targets are

---

[8]Lu et al. [8].

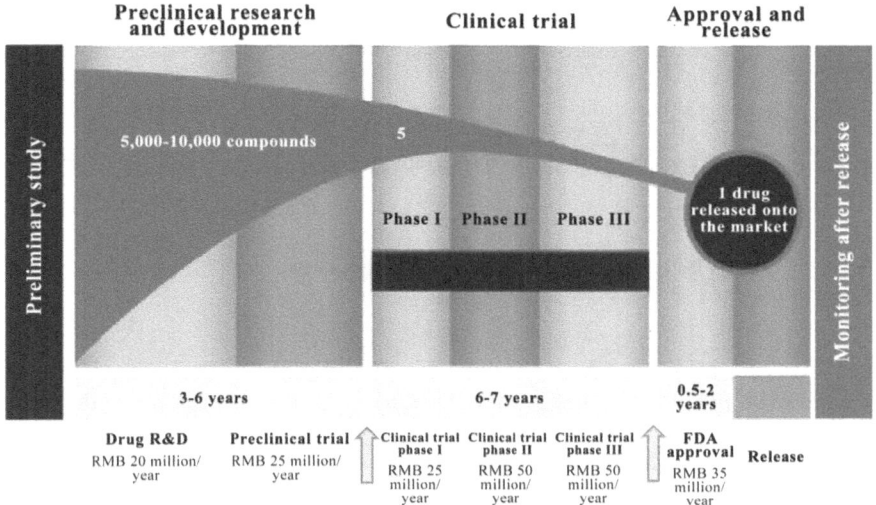

Fig. 3 Current drug R&D and launch process. *Source CEIBS: 3DHTS Innovation Analysis Report*

selected, a biological model should be established to screen and evaluate promising compounds. Compounds may be produced via a range of methods, most notably including extraction from plants or animals, organic synthesis and molecular modification. The activity of the selected compounds must be optimised and then evaluated via in vivo and in vitro studies, in order to acquire compounds that are most suitable as drug candidates.

The next stage is preclinical research. Pharmaceutical companies need to conduct lab studies and animal testing to identify the effects of drug candidates on the biological activity of tumour tissues, as well as to evaluate drug safety. It takes about three years to complete all of these studies.[9] After preclinical research, pharmaceutical companies should submit their Investigational New Drug (IND) Applications to the FDA for approval before beginning clinical trials.[10]

Clinical trials can also be divided into three phases with different scales and goals. Phase I lasts about one year, and requires 10–100 healthy volunteers. It focuses on drug safety issues such as safe dosage ranges and identifies the absorption, distribution, metabolism and excretion processes of an anti-cancer drug in the human body, as well as the duration of drug action. Phase II lasts about two years and needs 100–500 cancer patients to participate in case-control studies in order to evaluate drug effects. Phase III needs about three years, and usually involves 400-5000 patients in hospital. Doctors monitor the status of patients to

---

[9]Moffat et al. [9].
[10]Deng [10].

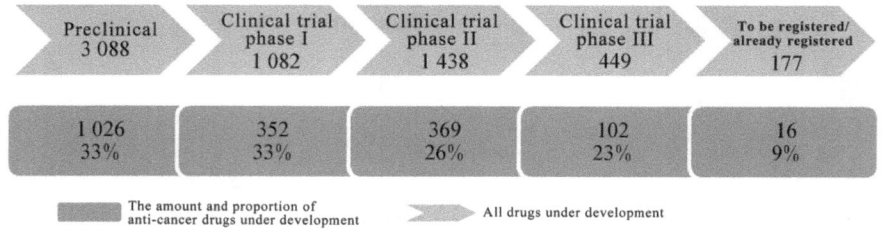

Fig. 4 2013 statistics of new anti-cancer drugs in development around the world. *Source* IMS: *Global oncology trend report*

identify drug effects and adverse reactions.[11] Then pharmaceutical companies need to analyse all the data acquired from clinical trials. If this data can prove the safety and effectiveness of the new drug, they can submit a New Drug Application (NDA) to the FDA. A typical set of NDA documents consists of 100,000 or more pages, and the review process usually takes 1–2 years. In the end, less than 10% of new drugs are able to get approval and enter the market.

There is also a fourth stage called post-market safety monitoring, during which pharmaceutical companies regularly submit reports to the FDA, including all reports on adverse reactions and certain quality control records. The FDA may require further research on some certain drugs to evaluate their long-term effects.

According to Roche, developing a new drug from primary lab studies to commercial sale in pharmacies takes 12 years, 6587 trials, 423 participating researchers,[12] and R&D expenses of USD 1.2 billion[13] on average. Among the new drugs in development around the world, nearly a third of those at the preclinical stage and undergoing phase I clinical trials are anti-cancer drugs,[14] but these only account for 9% of drugs in later phases (see Fig. 4). This data illustrates the great difficulty of developing anti-cancer drugs.

The Biopharmaceutical Industry Chain

From the perspective of drug product development, the anti-cancer drug industry chain is composed of upstream, midstream and downstream players. Upstream players focus on technological innovation and development; midstream players involve material separation and product processing; while downstream ones cover

---

[11]U.S. Food and Drug Administration, "The Drug Development Process" (24 June 2015), FDA, http://www.fda.gov/forpatients/approvals/drugs/default.htm, last accessed on 27 October 2015.

[12]Roche [11].

[13]Paul et al. [12].

[14]IMS Institute, "Global Oncology Trend Report: Innovation in Cancer Care and Implications for Health Systems", *IMS Reports*, May 2014.

marketing and planning, channel building and feedback system development. Due to strict technological requirements and numerous lab studies, the early-stage R&D expenses of anti-cancer drugs can be much higher than the direct cost of production. Therefore, the R&D departments of leading pharmaceutical companies, along with biotechnology companies, have become the core of the industry chain.[15]

In order to meet investment needs and reduce risk, multinational pharmaceutical enterprises, biotechnology companies and small-scale pharma firms usually build business alliances to provide joint investment for drug R&D. A common model is to put a small biotechnology company with strong professional skills in charge of technological development and innovation. Through cooperative development, the company can acquire the technologies to produce biological medicines, or the right of production. Pharmaceutical companies may also cooperate with competent contract research organisations (CRO) and subcontract studies that require advanced technologies to them, so as to reduce cost and ensure professional and efficient operation.

Analysis of Global Market Potential

As anti-cancer drug R&D has high requirements in terms of funding and technology, most branded drugs are produced by multinational pharmaceutical companies such as Roche, Novartis, AstraZeneca, Sanofi and Pfizer. The top ten anti-cancer drug producers accounted for 81% of the global market in 2010, and Roche, with its rich product range, contributed 42% of their total sales value.[16]

Multinational pharmaceutical companies are also increasing their investment in drug R&D. According to the Pharmaceutical Research and Manufacturers of America (PhRMA), R&D investment by American biopharmaceutical companies reached USD 67.4 billion in 2010. PhRMA members also increased their R&D investment by 6.5% and the ratio of R&D investment to sales value to 20.5%, which had remained around 19% from 2000 to 2010.[17]

Analysis of Chinese Market Potential

According to the *Blue Book on Chinese Pharmaceutical Market Development*, in 2010, Chinese hospitals consumed drugs worth a total of RMB 452 billion with a year-on-year growth of 22.5%, the retail pharmacy market was valued at RMB 173.9 billion with a year-on-year growth of 17%, and the community hospital and

---

[15]Wang [13].

[16]Li [14].

[17]"R&D Investment by U.S. Biopharmaceutical Companies Reached Record Levels in 2010", *PhRMA*, 15 March 2011.

rural clinic markets were valued at RMB 129.7 billion with a year-on-year growth of 27.9%. Due to population ageing, a new universal health care system, and the improved comprehensive payment capacity of Chinese people, China is expected to become the second-largest pharmaceutical market in the world by 2020.[18]

Driven by the huge Chinese market potential, multinational pharmaceutical companies have been increasing their investment in China. In March 2007, AstraZeneca announced the establishment of its Innovation Center China in Shanghai, focused on developing drugs for the most common diseases in Asia such as liver, stomach and lung cancers.[19] In October 2007, Roche opened its Pharma Development Center China in Shanghai. This is the first comprehensive clinical trial centre in China, and is able to meet all requirements for clinical trials. It also works with Chinese experts, scholars and professionals to explore innovative therapies.[20] In October 2008, Lilly launched its China R&D Headquarters in Shanghai to work on scientific research management and venture capital investment, and seek cooperation with Chinese research institutes and scientists.[21] In April 2009, Johnson & Johnson established an R&D centre in Shanghai as its Asia Pacific R&D Headquarters. The new centre pays special attention to developing drugs for common cancers, infectious diseases and metabolic disorders in Asia, and promotes open innovation by cooperating with colleges, universities and research institutes.[22] In November 2009, Novartis announced an investment of USD 1 billion over the next five years to establish the China Novartis Institute for BioMedical Research in Shanghai, which focuses on developing drugs for the most common cancers (such as stomach and liver cancers) and liver diseases (such as hepatitis and hepatic fibrosis) in Asian countries, including China.[23]

This means that local pharmaceutical companies in China now face greater pressure from an increasing number of their global counterparts entering the Chinese market. Due to their small scale and limited R&D investment, they mainly produce generic anti-cancer drugs, making them less competitive when facing imported brand-name drugs. According to Ding Jian, Academician of the Chinese Academy of Engineering and Director of the Shanghai Institute of Materia Medica, foreign pharmaceutical companies usually use 20% of their sales revenues for new drug development; however, this number is less than 2% on average in China.[24] One reason is that, China's regulated drug pricing system has reduced the profits of local pharmaceutical companies, making them unable to bear the risk of drug development failure. The other reason is that, the drug review process in China is

---

[18]CFDA Southern Medicine Economic Research Institute: *Blue Book on Chinese Pharmaceutical Market Development 2010*, 29 October 2010.
[19]Shi [15].
[20]Yu [16]
[21]Zhu [17].
[22]Yang [18].
[23]Xu [19].
[24]Liu [20].

time-consuming, which has negative effects on the innovation practice of local companies. In short, multiple factors, such as limited technology capacity and uneven industry development, have constrained the drug R&D capacity of local pharmaceutical companies in China. They cannot even compete with their global counterparts in terms of the quality of their generic drugs. To address this challenge, the China Food and Drug Administration (CFDA) began a quality consistency evaluation for 75 generic drugs in July 2013 in an effort to significantly improve drug quality and safety.[25] With this evaluation, as well as policies to encourage the development of first-time generic drugs, a range of low-quality and repetitive generic drugs will be phased out of the market, and local pharmaceutical companies will have to shift their focus to improved generic drugs and independent R&D.

The Start-Up Decision

The Early Years: Accumulating Knowledge

Dr. Xiong Lei studied at the Institute of Biochemistry and Cell Biology, SIBCB, CAS, and began researching oncogenesis mechanisms in 2000. After receiving his doctoral degree, he joined the newly founded company Abmart as the first manager of its Business Development Department. Abmart was founded in Shanghai, China in 2006 by Dr. Meng Xun, a biologist studied in the USA, and Dr. Chen Changzheng, a Stanford professor. The company specialises in the development and production of monoclonal antibodies. After Dr. Xiong Lei joined the company, he took charge of customer services, technical support, sales and marketing, and led his team to increase the sales volume by 150% during his time there.

Nevertheless, Xiong Lei wanted to get more involved in scientific research, so he left Abmart in 2008 after a short time working there, and began his postdoctoral research at the University of Zurich in Switzerland. There he focused on high-throughput target screening with RNA interference (RNAi) technology, engaging in drug target development and verification guided by a strategy of "systems biology". During his studies in Switzerland, Xiong Lei also completed a short-term MBA training course at the Swiss Federal Institute of Technology in Lausanne.

Forecast of Market Trends

In 2009, during his postdoctoral research in Switzerland, Dr. Xiong Lei came across a report from Applied Biosystems (ABI) by chance. Based on analysis, the report

---

[25]Wang [21].

predicted that the cost of human whole genome sequencing (WGS) for an individual would drop to USD 1000 over the next five years, while Xiong Lei's research institute was spending around EUR 100,000 on procurement at the time. Although this prediction was not inconceivable to Xiong Lei, as he had heard about such targets during his doctoral research in 2006 and was in regular communication with friends who worked in related areas, with the continuing development of computing and storage technology, Xiong Lei and his friends eventually believed that "precision treatment" supported by big data would come true. "Because researchers should look to the future, not to the past or the present. As business-people we should sometimes cater to present needs, but as researchers we should be forward-looking and focus on cutting-edge topics. Being trained to look to the future, we began to wonder what would happen if the genome sequencing cost were to drop to USD 1000 in five or six years' time, and how it could change the world. It was very exciting to think about it."

Meanwhile, Xiong Lei saw huge potential in the Chinese market. He was deeply inspired by the book *China's Megatrends*,[26] by John Naisbitt, and even bought his friends copies of the book in order to share this inspiration with them. The book predicted that China would become the world's largest economy—a truly developed country—and analysed the reasons for such a trend in detail.

After some discussion, Xiong Lei and his friends came to the conclusion that the pharmaceutical sector would follow the same trend: "When China transforms into a highly developed country, the situation of its pharmaceutical sector will change correspondingly. A stronger China will not end up producing generic drugs all the time and relying exclusively on technologies developed by other countries. Things simply cannot happen that way. Innovative enterprises will be founded. If not by us, then others will do it. We are probably among the first batches to step forward and strive to ride the trend. Though we don't know yet whether or not we can keep the leading position, we will not regret our decision."

Back to the Homeland

On Christmas Eve, 24 December 2009, Dr. Xiong Lei made up his mind to start up his own business back in China. He wanted to fully leverage the rich clinical cancer resources available in China and establish a personalised drug development platform using systems biology, setting milestones for Chinese drug development.

Xiong Lei spent a year organising his team based on genome information guided drug development industry chain, through which he gathered resources for complete drug target screening as well as gene function research, and prepared for the creation of biological and medical databases.

---

[26][US] Naisbitt [22].

On 26 November 2010, Xiong Lei stopped his postdoctoral research and went back to China to start up his own business.

The Start-Up's Focus

Individualised Precision Treatment

In traditional anti-cancer drug treatment, chemotherapy remedies are chosen based on where the cancer appeared. However, as cancer is caused by gene mutation, and where the mutated site occurs is random and selected through evolution, mutation types vary from individual to individual.[27] Therefore, even when the same amount and same kinds of targeted drugs and the same therapy are used for cancer of the same tissue, clinical effects and adverse drug reactions may vary greatly. This is believed to be the main reason that the effectiveness of anti-cancer drugs is less than 30%.[28]

Since the human genome was mapped in 2003,[29] people have gradually formed a deeper understanding of genes, and paid more attention to pharmacogenomics, which studies the role of gene mutation in drug response.[30] But traditional pharmacogenomics research mainly studies how genetic polymorphism in the blood influences the efficiency of drug metabolism in different individuals. For cancer, however, the tissue consists of unique somatic cell genomes where mutation may occur at any sites, so it is necessary to study how somatic mutation within the cancer genomes influences drug response.

Dr. Xiong Lei and other researchers began to contemplate whether it was possible to tailor medicine therapy based on the pathogenic genes of different patients, in other words, to achieve "personalised precision therapy".[31]

Large-Scale Drug Screening Model

To achieve "personalised precision therapy", the range of types of anti-cancer drugs on the market ought to be greatly expanded. However, under the traditional drug development mechanism, phase II clinical trials are carried out randomly on cancer patients with various genotypes and as a result, anti-cancer drugs targeting a small

---

[27]Chen [23], Zhou and Liu [24].

[28]Wang and Baorui [25].

[29]"2003 Release: International Consortium Completes HGP", *National Human Genome Research Institute*, 14 April 2003.

[30]Luo [26], Sun et al. [27].

[31]Calvo et al. [28].

group of genotypes (for example, 5% of patients) cannot achieve a 20% effective rate, and thus cannot be approved by FDA to enter the market. Therefore, pharmaceutical companies need to search for target cancer patients as trial subjects for their drugs before large-scale clinical trials in order to significantly increase the efficiency of the clinical trials. For example, if a lung cancer drug is estimated to have an effect in 5% of patients, and 100 subjects are required for a phase II clinical trial, it is recommended to first select 100 people from 5000 through clinical diagnostic tests, an approach which is accepted and promoted by FDA. The FDA even expedited pathways for this kind of clinical trials with regard to potentially high effectiveness drugs, or for "breakthrough therapy"[32]—once a drug is designated as a breakthrough therapy, its clinical trial will be put on a pathway to accelerated approval.

To match cancer patients with the appropriate drugs, a cancer drug screening model is needed to identify the "gene-drug interaction". This is also a core process in initial drug development. Cancer cell lines are used to identify genetic biomarkers, but creating them is originally an extremely difficult process with a low success rate. This means that a large number of cancer samples are wasted, resulting in cost issues even for multinational pharmaceutical companies. However, there are rich clinical cancer resources in China, and if they are fully exploited, cadaver tissue banks can be transformed into cell banks, allowing for high-throughput drug screening and the accumulation of huge amounts of "drug-gene" data, thus providing first-hand big data for drug biomarkers (matching drugs to genes).

Next-Generation Sequencing (NGS)

Undeniably, even if there are enough cancer cell samples, it is not easy to match drugs to genes due to the relatively large amount of data. Human somatic cells have 23 pairs of chromosomes, with a number of genes linearly arranged on each chromosome.[33] Preliminary analysis of the human genome map, conducted by scientists from many different countries, has revealed that humans have around 20,000–25,000 genes in total.[34] Supposing the total number of genes is 25,000, this makes for over 300 million potential combinations of two genes, and over 2.6 trillion combinations of three, so finding the correlation between genes and diseases is like finding the power switch in a complex circuit diagram[35] (see Fig. 5).

As a result, it usually takes years of research and enormous capital to identify a group of cancer biomarkers. Six countries spent USD 3 billion on the first genome sequencing in human history, and it took about ten years. Even after 454 Life

---

[32]Sherman et al. [29], Subramanian et al. [30].

[33]"2003 Release: International Consortium Completes HGP", *National Human Genome Research Institute*, 14 April 2003.

[34]Gangwar and Worabo [31].

[35]McLeod and Evans [32].

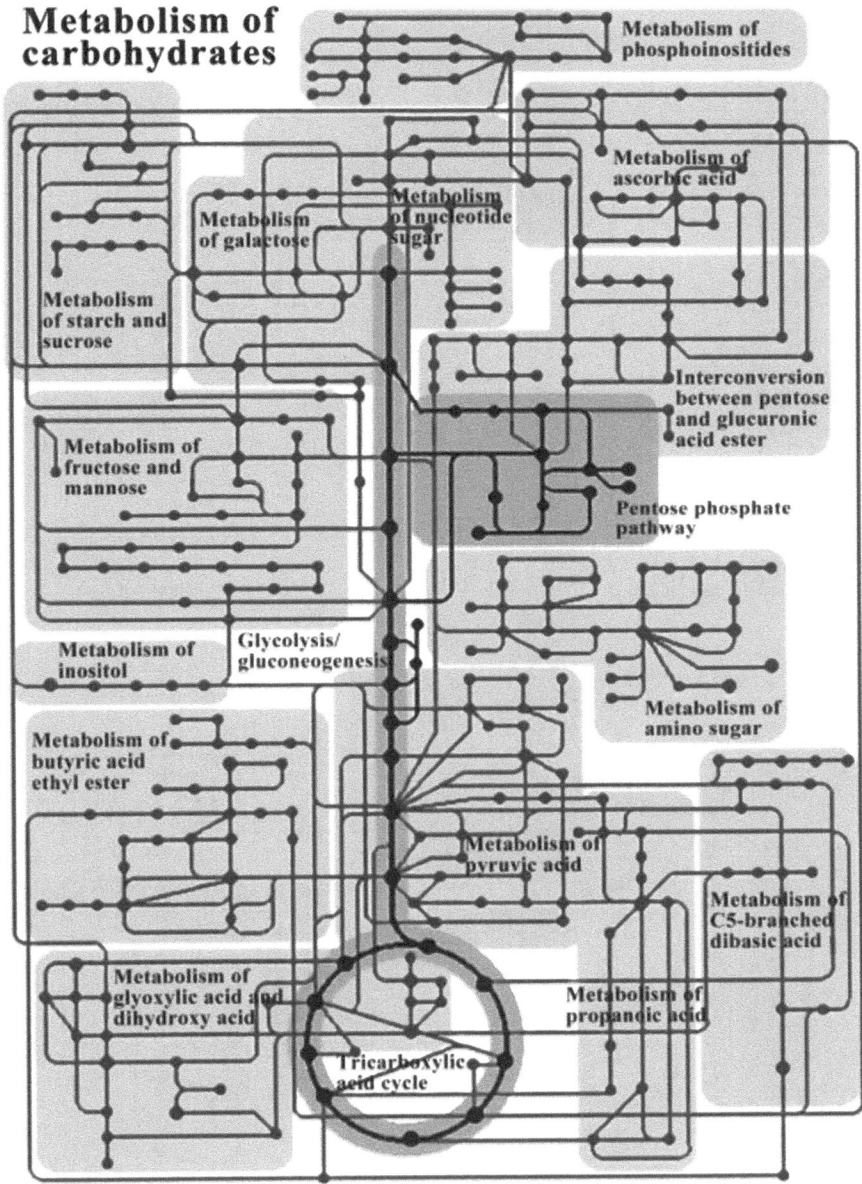

Fig. 5 The difficulty of identifying drug-disease relationships. *Source* 3DMed internal materials

Sciences released a next-generation sequencing (NGS) instrument at the end of 2005, the cost of personalised genome sequencing was only reduced to several million RMB.

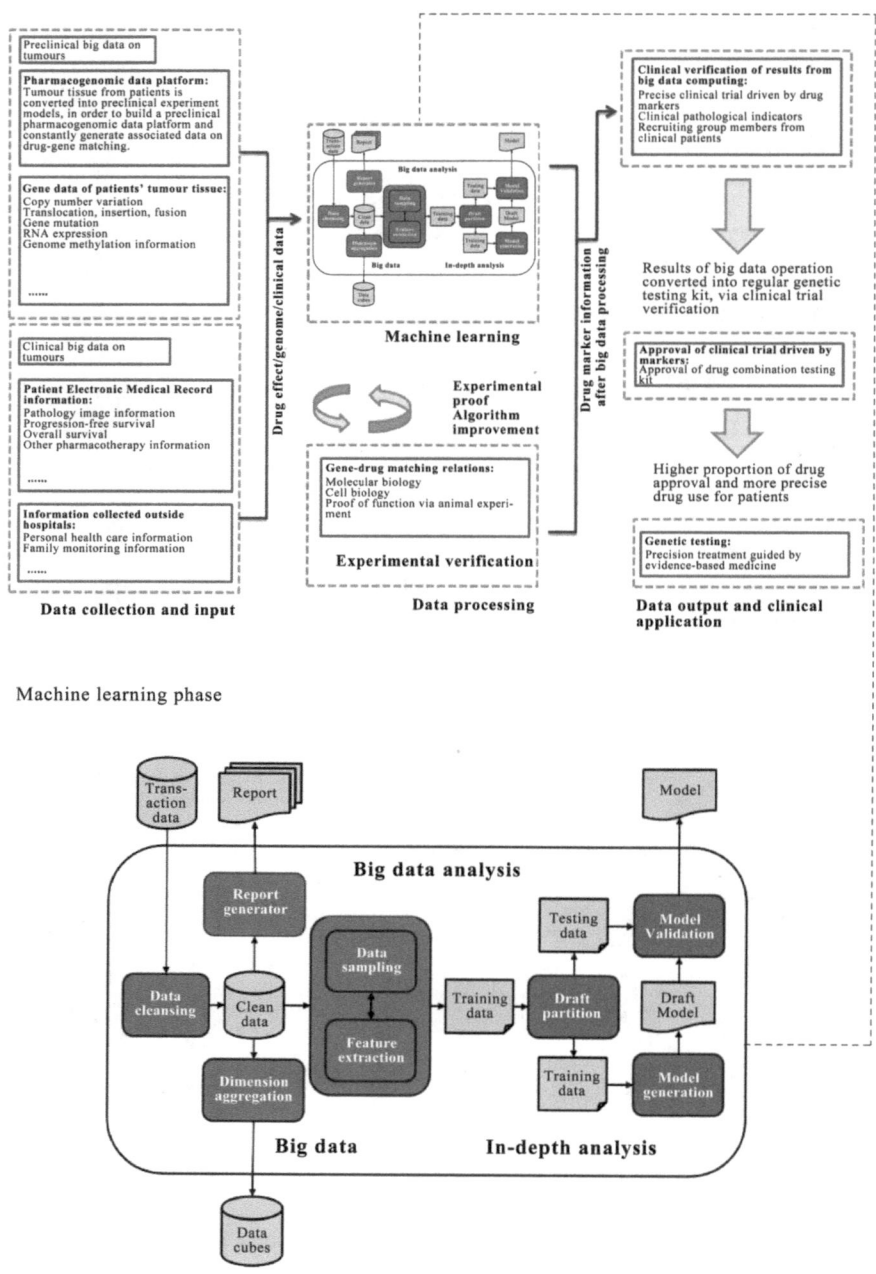

Machine learning phase

Fig. 6 Biomedical science big data flowchart. *Source* 3DMed internal materials

With global biotechnology developing, NGS technology has revolutionised genome sequencing, bringing many benefits compared to the first generation sequencing methods. NGS technology is characterised by large-scale and high-throughput sequencing capabilities compared to traditional antimicrobial susceptibility testing technology. By virtue of automated operation systems, sensitive and fast detection instruments as well as high-speed data analysis computers, NGS technology allows hundreds of thousands to millions of DNA molecules to be sequenced simultaneously,[36] dramatically driving down labour and material costs.

Big Data Processing

High-throughput instruments can detect specific changes in DNA molecules in a short time and produce a large amount of information. For example, the Illumina HiSeq 2500 whole genome sequencing instrument can completely sequence an entire genome in just one day. It can produce 120 GB data in 27 h, or 600 GB data during one time of standard operation, which is far higher than the average sequencing output a few years ago.[37] High-throughput methods generate huge amounts of "big data", and the capability to process such data is thereupon needed. In short, drug-gene relations can only be revealed by integrating cancer genome data with high-throughput drug data.

Around 2010, with the emergence of cloud computing and cloud storage technologies, people have become able to process massive amounts of information. Thus, Dr. Xiong Lei planned to integrate oncogenomics, complex data computing platforms and high-throughput drug screening technology to solve problems in personalised precision anti-cancer drug R&D (see Fig. 6).

The Start-Up Process

Founding the Company

After integrating all the thoughts and ideas, Xiong Lei registered and founded 3D Medicines Corporation in 2011, using all of his savings—RMB 500,000. The "3D" in the company's name stands for the three Ds of Diagnostics, Drugs and Development.

Due to limited funding, the company could not afford to set up its own lab to begin with, and had to rent a lab in Jinshan District, 10 kilometres away from the Shanghai urban area. Every week, employees had to commute from Jinshan to the

---

[36]Han and Yang [33].

[37]Liu [34], Herper [35].

city to do business. The company made its first significant profit within half a year by providing outsourced science and research services. In July 2011, it rented a 150 m^2 factory in Shanghai Juke Biotech Park in the Xuhui District, where it established its own lab and started to purchase instruments and facilities. After two years of development, the company expanded its scale and moved to Pujiang Hi-Tech Park, an extension of Shanghai Caohejing Hi-Tech Park with energy, electronic information and biomedicine as its core industries. 3DMed occupies a 2400 m^2 lab within this industrial park.

The Entrepreneurial Team

Right after the company was founded, the entrepreneurial team consisted of eight members (see Fig. 7). With a similar academic background of CEO Dr. Xiong Lei, two other members of the management team—Technical Director Dr. Andy Xie and Gene Function Manager Dr. Li Fengqing—also graduated from the Institute of Biochemistry and Cell Biology, SIBCB, CAS. Database Business Unit Manager Zhang Qingzhou graduated from the University of Manchester, and had worked at Abmart before he entered 3DMed, while Sales Manager Dr. Li Huaguang and Business Development Department Manager Dr. Fang Qiangyi graduated from the CAS-MPG lab and the Shanghai Research Center of Biotechnology of CAS respectively. The team was later joined by An Yinghui and Li Xiaofang. An Yinghui graduated from East China Normal University, and had previously worked in sales at Abmart under Dr. Xiong Lei, while Li Xiaofang had graduated from CAS and followed Xiong Lei in scientific research.

Core team (significantly changed already)

Dr. Xiong Lei, Chief Executive Officer. Graduate from Institute of Biochemistry and Cell Biology, SIBS, CAS, with a doctoral degree, majoring in cancer drug resistance and signal pathway; 14 years of experience in oncogenesis; appointed as the first business development manager in Abmart, responsible for customer services, technical support, sales and

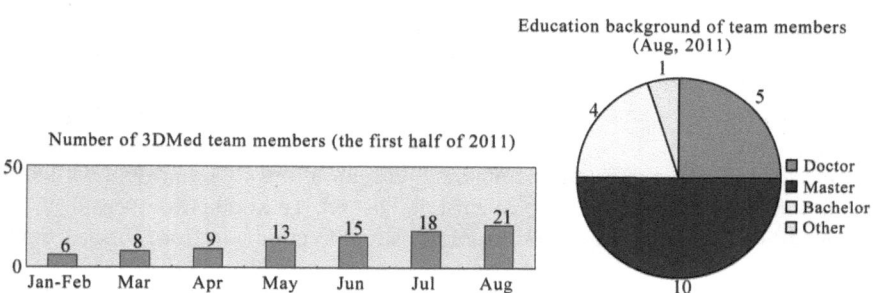

Fig. 7 The 3DMed team during the company's initial start-up phase (2011–2012). *Source* 3DMed internal materials

marketing, successfully led the team to achieve a 150% sales volume increase; engaged in high throughput screening (HTS) during his postdoctoral years in University of Zurich, where he mastered HTS RNAi screening technology, mainly in systemic cell apoptosis, autophagy, endocytosis, and ageing; also finished a short-term MBA training course at the Swiss Federal Institute of Technology in Lausanne. Responsible for defining strategies, creating corporate culture, designing genetics-related database products, and daily operations management during the initial phase of 3DMed.

Dr. Xie Zhenghua, Chief Technology Officer. Graduate from Institute of Biochemistry and Cell Biology, SIBS, CAS, with a doctoral degree, majoring in epigenetics. Spent postdoctoral years at the University of Rochester, where he mastered transgenic mice and gene knock-out mice technology and successfully applied this technology to diabetes model research. Responsible for building HT libraries, HTS, as well as building transgenic mice animal model platforms based on shRNA libraries during the initial phase of 3DMed. Currently responsible for building the company's cell model construction platforms.

Dr. Li Fengqing, Gene Function Manager. Graduate from Institute of Biochemistry and Cell Biology, SIBS, CAS, with a doctoral degree, majoring on control of gene expression, mentored by Hong Guofan (a member of Chinese Academy of Sciences). Spent postdoctoral years at the Institute for Nutritional Sciences, SIBS, CAS, majoring on animal models of metabolic diseases; 10 years of experience in molecular biology and gene functions, especially in control of gene expression. Responsible for annotating information in the genetic association database as well as conducting research on gene expression during the initial phase of 3DMed.

Zhang Qingzhou, Database BU Manager. Graduate from the University of Manchester with a master's degree, majoring on bioinformatics; built first database for production management and customer relations at Abmart; built and released an online database of protease cleavage sites during postgraduate studies (NickPred Database). Responsible for building gene expression database and gene-disease association database during the initial phase of 3DMed.

Dr. Li Guanghua, Sales Manager. Graduate from CAS Max Planck Institute (jointly created by China and Germany), with a doctoral degree, majoring on epigenetics; rich experience in RNAi mechanism of action; first discovered epigenetics changes caused by RNAi in drosophila; spent postdoctoral years at the Institute of Biochemistry and Cell Biology, SIBS, CAS, mentored by Liu Xinheng (a member of Chinese Academy of Sciences), majoring on cancer gene therapy, mainly regarding how to use RNAi technology to inhibit gene expression in cancer treatment. Responsible for expansion of sales and marketing in HTS-related scientific research business during the initial phase of 3DMed.

Dr. Fang Qiangyi, Manager of Business Development. Graduate from Bioengineering Research Center of the Chinese Academy of Sciences, Shanghai, with a doctoral degree, majoring on large-scale expression of eukaryotic recombinant proteins in mammalian cells; experience in the expression of several dozen large-scale eukaryotic proteins; consecutively responsible for leading technical polyclonal and monoclonal antibody R&D in Abmart, development of polyclonal antibody products, and sales of these products to industrial customers; rich experience in development of technologies for mass production. Responsible for project management and assisting in the management of library production.

By August 2011, the 3DMed team had grown from eight members to 20, after which it remained more or less the same size for a time. In 2012, the team was expanded to about 40 members, with more employees hired for shRNA library creation, drug target screening and sales forces. They did some basic work for outsourcing of science and research services, which contributed to the company's capital accumulation.

Accumulating Resources

After analysing the competitiveness of 3DMed, Xiong Lei found that the company faced two main problems in its initial stage: a lack of funding and a lack of cancer sample resources.

To accumulate funds initially, 3DMed adopted a "group buying" model for technical services, and bought in large numbers of research-related service contracts at relatively low prices. Afterwards, under the conditions of big contracts, 3DMed negotiated with upstream suppliers, winning the terms of purchasing the raw materials and equipment at RMB 9 million with one-year payment cycle. Since 3DMed was allowed to pay by instalments, it could perform service contracts as non-core business using the instruments and raw materials to bring in cash, which was then used to pay for the instruments. In this way, it leveraged the accumulated fixed assets of merely RMB 1 million to launch a shRNA library creation business worth RMB 10 million. A shRNA library is a commercial application based on large-scale RNAi screening technology. Researchers can simply search for the shRNA of a specific gene in the library, saving them the trouble of siRNA design, synthesis and verification processes (see Table 1). This kind of service business bought Xiong Lei time to devise a strategy to ensure the start-up's survival and the ability to finance its development.

As for cancer sample resources, he decided to cooperate with hospitals by providing them with shRNA services for a year or two. During this period, 3DMed mainly cooperated with research hospitals that specialised in translational medicine

Table 1 3DMed libraries in the early stages (2011–2012, before A-round financing)

30,000 gene switches (library products)		
	siRNA library	shRNA library
Manufacturing	Chemosynthesis	Raw materials of chemosynthesis +biosynthesis
Animal model	Not reusable	Can be used in animals
Minimum technical requirements	Low	High
Manufacturing countries	USA, China, South Korea, etc.	USA
Number of manufacturers	>6 (in China)	3 (globally)
Library product pricing	Stuck in price war	Has monopoly price
	Sigma	3DMed
	RMB 2500/iRNA	RMB 150/cost
RNAi screening	High cost and small scope of application	Low cost and wide scope of application

Source 3DMed internal materials

research, such as Zhongshan Hospital, affiliated to Fudan University, and Renji Hospital, affiliated to the School of Medicine of Shanghai Jiao Tong University. Translational medicine research deals with how basic research findings can be applied to clinical practice. For example, the "targeted drug susceptibility testing" offered by 3DMed can enable the personalised application of drugs, and its "drug target research" contributes to reveal new anti-cancer drug targets and guide drug development.

Though these technologies earned 3DMed some revenue and data resources, Xiong Lei wanted to focus on more than just these fields, saying that "These are common technologies, and our implementations are only slightly more complex than those of other companies". At that time, 3DMed was struggling for survival. "We have to wait and build up our networks, and turn to personalised anti-cancer R&D as soon as we have raised a large amount of initial capital."

Angel Round of Financing

According to Xiong Lei's original plan, 3DMed would attract investment after it accomplished the expected profit goal in the first three years after its founding. However, it accomplished the goal sooner than expected. Thus, the year 2012 became a milestone in Xiong Lei's career and the development of 3DMed. In May 2012, Xiong Lei joined the "1st Future Leaders' Boot Camp" Leadership Development Programme at the China-Europe International Business School (CEIBS). In August 2012, 3DMed completed its first round of financing and raised RMB 10 million in angel funding.

As soon as the financing arrived, Xiong Lei began implementing his plan. According to the plan, 3DMed aimed to create cancer cell line models on a large scale in cooperation with hospitals and integrate genome sequencing data with high-throughput drug screening data for personalised drug development (see Fig. 8).

Fig. 8 3DMed's capital plan after A-round financing (after A-round financing in 2012). *Source* 3DMed internal materials

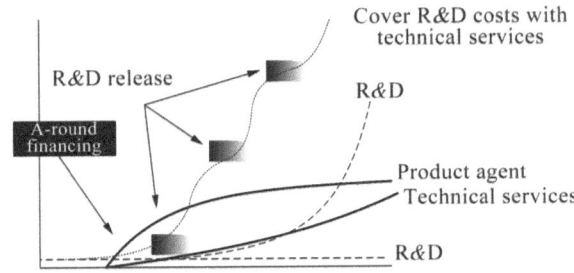

Technological Innovation

Though there are hardly any direct competitors in China at the time, and it could earn high profits from conducting precision diagnostics business with drug susceptibility sequencing technology, after receiving the angle round of financing, the 3DMed management team decided to upgrade its equipment to NGS technology. This required a huge input of capital and talent, and would cause high R&D costs in the initial phase. But once the technology was upgraded, a clearer, modular process would lead to a decrease in marginal costs, from RMB 100,000 down to just RMB 15,000. Though some profit margins might be sacrificed, the 3DMed management team believed in the long-term strategic value that such a technological upgrade would bring. As 3DMed Public Relations Director Li Yishi said, "The upgrade would expand our target customer segment and raise the technical barrier for potential competitors. At the same time, we believe that we are in the era of big data, where collecting user information is more important than gaining short-term profit. Only by collecting massive amounts of cancer genome data, can we unleash the potential of future business operations."

Their strategic direction was recognised by angel investors.

The Business Value of 3DMed

Providing New Drug R&D Services

In the second half of 2012, 3DMed's core service shifted from the RNAi library service to drug marker screening and R&D using NGS technology.

3DMed cultured cancer cells in vitro and developed cell models for preclinical trials of anti-cancer drugs to figure out the genotypes that the drugs could apply to. Analysis of the association between genotypes and efficacy of drugs, using large amounts of data, could demonstrate whether the drug was specific to a certain genotype. Preclinical screening of patients helped pharmaceutical companies find people specific to the drug who could participate in the trial, greatly increasing the trial efficiency (see Fig. 9). It also helped pharmaceutical companies reduce costs significantly, because although the cost of gene sequencing had fallen to USD 1000 per capita, the cost of clinical trials could be as high as USD 100,000–200,000 per patient. And what's more, without genetic biomarkers, most clinical trials failed anyway.

Screening drug biomarkers not only greatly increased the chance of new drugs being approved by FDA, but also enabled drugs whose clinical trials had failed to find specific target patients and be redeveloped. In terms of specific cooperation, 3DMed firstly aimed to sign agreements with large innovative global pharmaceutical companies (such as Roche, Novartis, Johnson & Johnson, and Eli Lilly) to redevelop their drugs whose clinical trials failed by finding the precise target

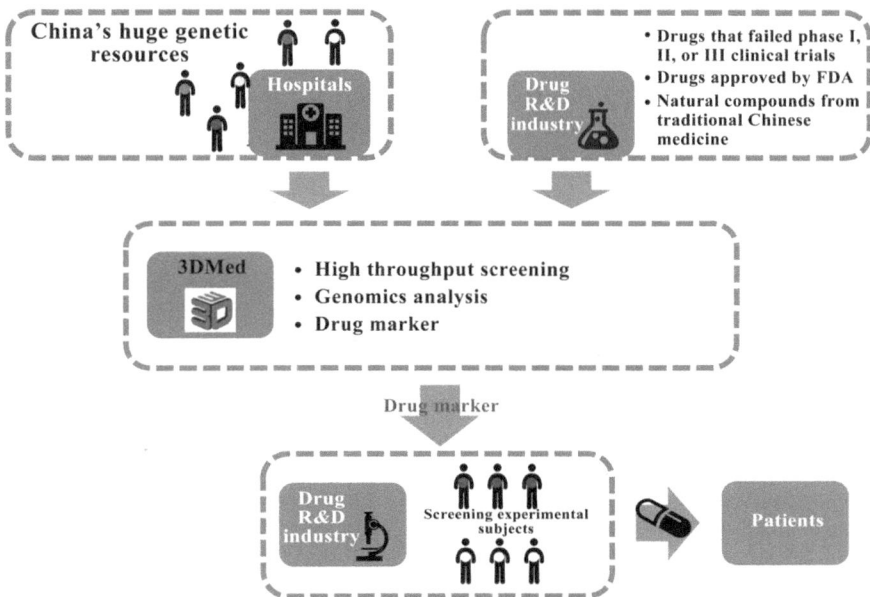

Fig. 9 3DMed "drug-gene" matching association study (after A-round financing). *Source* CEIBS: 3DHTS Innovation Analysis Report

patients and restarting clinical trials. Secondly, 3DMed was able to find specific biomarkers using platform screening, and jointly developed personalised precision drugs with other new drug development companies.

Building a Drug R&D Platform

3DMed's goal was not simply to provide CRO services for new drugs. Instead, 3DMed aimed to build an R&D platform for precision anti-cancer drugs.

3DMed started with liver cancer, one of the most prevalent cancers in China, by collecting samples of cancer cells from various top-tier (class 3A) hospitals and culturing these cells in laboratories. Once the number of cancer cell samples of one genotype exceeded a certain amount, 3DMed could develop personalised precision drugs with specific biomarkers. Within three years of its founding, 3DMed had established the largest database of liver cancer cells in the world, and the amount of data continued to increase. Xiong Lei aimed to increase the number of samples to 10,000 and develop a favourable business ecosystem in the biomedical industry by opening up data source.

Changing Clinical Diagnostics Models

Innovations in the drug R&D model would spurr the launch of more and more anti-cancer drugs in the future. Xiong Lei predicted that in the coming 10–20 years, "Theoretically, there will be 20–30 precision drugs focusing on different targets for each type of cancer, maybe even 50 drugs for one cancer". By then, patients were able to undergo WGS to identify their specific genotype, so as to determine the appropriate drug to realise personalised precision therapy.

Therefore, 3DMed not only applied drug R&D technologies, but also had an effect on the clinical diagnostic model in the downstream value chain. Traditional "companion diagnostics" only diagnosed a single gene and cost RMB 2000 in China, so diagnosing 10 genes cost RMB 20,000 in sum. The new WGS technology adopted by 3DMed, on the other hand, would only cost RMB 5000–6000 in future.

As the number of drug types increased, single-gene sequencing technology could not meet large-scale diagnosis needs. Xiong Lei believed that traditional single-gene sequencing would be phased out in the future, with significant consequences for the whole industry. "Big data changed the models of medication, diagnostics, and drug R&D. It served as a source that facilitated a series of chain reactions and changed different things in different phases. It changed drug R&D first and will change diagnostics models in the future. It has already changed diagnostics to some extent, with several new drugs put into clinical practice, but this only covers a few genes. For instance, there are currently only two to three drugs for one specific type of cancer. Patients might therefore think that adopting WGS is a bad deal, because single-gene sequencing technology only costs RMB 5000–6000 for two or three genes, while the WGS technology also costs RMB 5000–6000. If just one more drug became available, things would be different. The old sequencing technology will be phased out if one more drug appears for some cancers, because the old technology cannot save cost, while the new technology saves more and more as the number of genes involved increases. The cost remains the same for 10 genes, 20 genes, or even 200 genes."

By guiding personalised medication and charging fees for tests and evaluations, 3DMed had expanded its business scope from drug R&D to personalised precise diagnostics by 2013, and it planned to gradually get involved in the entire industry chain as time went on.

Future Challenges Facing 3DMed

Guided by its strategy blueprint, 3DMed began to follow its anticipated development path. But was it the right moment? What difficulties needed to be addressed? After reflecting and thinking over these questions, the management team of 3DMed concluded that there were three main aspects that posed challenges.

Sustainable and High R&D Investment

How to gain recognition from the capital market is often the first challenge faced by innovative start-ups in emerging industries. In May 2014, 3DMed completed A-round financing, receiving over 10 million RMB from six institutes. Two investors engaged in the angel round also invested in the A-round, mainly in the form of medical fund. 3DMed used this capital to purchase equipment and facilities, expand its primary cell line platform, and recruit more staff. Given the huge cost of R&D, Xiong Lei believed that 3DMed needed to obtain more resources in the future.

Undeniably, it's difficult to precisely predict 3DMed's future cash flow situation, because it is in an emerging industry with few mature enterprises to learn from. Additionally, most investors are tending to be profit driven. Therefore, 3DMed must address the challenges of attracting angel investors and venture capital firms, and maintaining a sound financial position despite high R&D costs.

Putting a Team Together After Strategic Transformation

In light of 3DMed's strategic transformation, its middle and senior management needed to have leadership competency as well as expertise in various subjects, including biology, new drug development, clinical medicine and information technology. However, as 3DMed aimed to innovate in an area that had been seldom explored in China, there was little suitable talent to be found in the market that matched the company's requirements. Xiong Lei said that 3DMed found it very difficult to recruit talented team members in the beginning. "Even headhunters specialising in biology related field couldn't select a single resume that met our basic requirements within six months. This may be unthinkable in other industries, but innovative enterprises always face this challenge."

Because such interdisciplinary talent was so hard to find, 3DMed turned to internal training. However, even doctoral and master students from top colleges both within China and abroad often found the training to be too intense for them. More than 60% of 3DMed's original team members left the company within two years of its founding. What's more, after completing A-round financing in May, 2014, 3DMed started to build a gene sequencing and anti-cancer precision drug platform. Some members of the management team could not adapt to the "re-start-up", both in terms of mindset and capability, and three core members left 3DMed. 3DMed summarised the features of these employees and improved its requirements for middle and senior management as well as technicians. They did not just expected to have the relevant expertise and learning capacity – they also needed to have some achievements in the field already under their belts, and share the entrepreneurial dream. 3DMed stuck to the principle of "attracting top talent with a promising vision and excellent wealth sharing mechanism" and offered stock

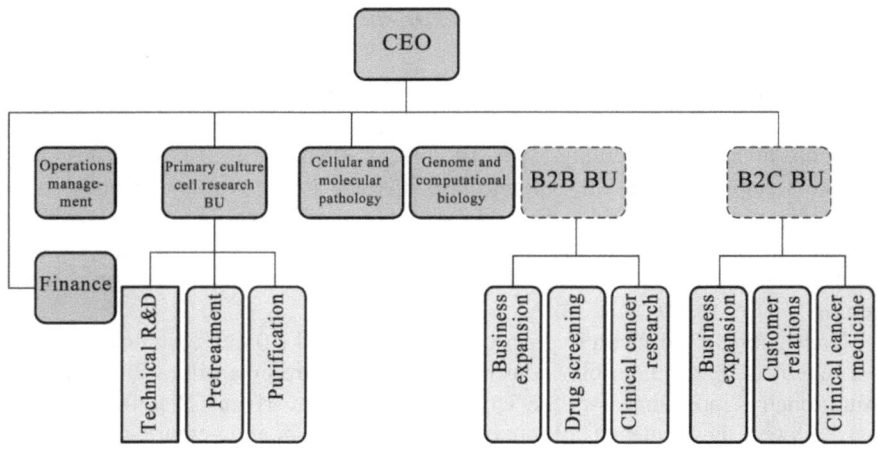

Fig. 10 3DMed organisational chart (2013). *Source* 3DMed internal materials

options as an incentive. It aimed to become a company carrying out employee stock ownership plan.

In addition to operation management and research teams, there were also some process-oriented manufacturing departments (see Fig. 10). Routine tasks such as cell model production required process stability. So 3DMed cooperated with vocational-technical schools for campus recruiting. In the beginning, up to 60–70% of recruited students were ultimately dropped due to their lack of technical skills, resulting in a significant waste of R&D investment. Later, 3DMed changed its recruiting model and pre-screened students just as it pre-screened drugs. "We offer a targeted training course, and I will teach the students in the sophomore year monthly. I will tell them what challenges they are going to encounter in 3DMed, both mentally and physically: 'It's going to be hard, but it's a good opportunity to learn. Your job will be focused on research and trials, so you have to really love this kind of work.' Most students are eliminated in the initial stage, while only 20 out 100 stay. But the turnover rate is only 10–20% once these 20 students are working as interns in 3DMed. After students start full-time work with 3DMed, they seldom resign. This is a valuable lesson that over two years of experience has taught us. It's similar to big data analytics. We have to carefully select certain data, or students, in the beginning in order to get the results we want."

The number of employees working at 3DMed has grown to more than 100, and Xiong Lei hopes that the number will reach 500–1000 in the next few years, and that 3DMed will build a marketing team covering the business fields of precision therapy and diagnostics. Will such recruiting and training models let 3DMed build its dream team? Can 3DMed's organisational structure and management model adapt to such goals? Time will tell.

Cultivating Customers in a Changing Industry

Pioneers in an industry not only need to grasp early opportunities, but also must bear the brunt of risks. Xiong Lei once said that "personalised medication moves too many people's cheese", alluding to the famous motivational parable by Spencer Johnson.

Traditional pharmaceutical companies have a mixed attitude towards personalised therapy. When a drug developed by such companies manages to obtain FDA approval, the companies often want this drug to be applicable to a broader range of people. They may find personalised therapy undesirable because it does not bring in as much profit. However, when the drug R&D fails, companies desperately hope that personalised therapy can help to identify specific target patients for the drug, so that they can continue to conduct clinical trials and launch the drug again. But once the drug is launched, companies hope that more patients will buy it, which goes against the intention of personalised therapy. To strike the right balance between precision therapy and business interests, drugs targeting 6–7% of the whole population are already appearing on international market and selling well. 3DMed, however, wants to develop precision drugs that target over 1–2% of the whole population, in line with the conclusions of relevant research by MIT.[38]

For doctors in hospitals and other medical units, personalised therapy may completely change disease diagnostics and medication. Are doctors willing to break with their reliance on old therapy methods, and adopt more complex but more precise therapies? Xiong Lei was optimistic in this regard. "During exploration of new therapy methods, we found that there will always be some doctors who care about patients' health above all else. They chose the old therapy because they had no choice; when we offered them the new therapy, they were uncertain whether to use it. Different doctors reacted differently, and we felt their hesitation. But I believe that this was a good phenomenon in itself, because they did not even have anything to be hesitant about before." As the personalised medical industry gradually matures, Xiong Lei expects that the price of personalised gene sequencing will reduce to RMB 10,000 in the future, which is affordable for individual consumers.

As the reform of China's health care system deepens, 3DMed must grasp the opportunity of moving away from traditional therapy towards personalised therapy, sticking to its business values while facing various new challenges. This is not a simple task. With A-round funding in place, 3DMed will enter a new stage of development. Are Xiong Lei and his team ready for this? How can 3DMed develop its precision therapy business model based on WGS big data?

---

[38]Sharma et al. [36].

Case Analysis I

Grasp Consumer Demand, Stick to Innovative Ideas

Fan Xiaojun

"Precision", "personalised", "customised", "targeted (therapy)"... Most oncologists are very familiar with these terms, which are often used to describe the management model of developing a treatment plan based on the patient's individual features and the genetic characteristics of the tumour. Precision therapy in the medical industry is enabled by technological development and the application of big data.

This case clearly shows that enterprises must truly understand and grasp market demand in order to develop products and conduct marketing activities. To identify and grasp consumer demand, enterprises must conduct comprehensive and scientific market research. Anti-cancer drugs are now a hot spot for innovative drug R&D both domestically and abroad. Substantial capital is flowing into cancer research, and numerous new products are entering the clinical trial stage. Most domestic R&D staff has been engaged in development of generic drugs for a long time, while little effort has gone into the clinical development of innovative drugs. However, relevant research has helped 3DMed and Xiong Lei realise the great potential of this area. Xiong Lei conducts research on personalised precision therapy based on his own expertise and research experience, and manages to meet consumer demand by reducing the cost and sale price of the therapy, so that consumers can afford it and are willing to pay for it. 3DMed always strives to gain an in-depth understanding of consumer demand, and adapt to this demand when developing personalised anti-cancer precision drugs. This fully demonstrates the essence of modern marketing, i.e. "providing appropriate products to appropriate consumers at an appropriate price, at the appropriate time and place", as well as the fundamental spirit of modern marketing, namely focusing on consumer demand and striving to earn long-term and reasonable profit while meeting this demand.

A good business model can seamlessly integrate technological innovation, product innovation, and service innovation, bind the interests of various links in the industrial chain together, and continuously promote development of the industrial competition model and progress of the economy. During this process, technological innovation and business model innovation complement each other and achieve coordinated development. In a sense, technological innovation is a prerequisite for business model innovation. More advanced technology enables changes in the profit model and source of profit. In terms of the drug development process, the upstream part of the anti-cancer drug industrial chain involves key technological innovation and development, while the midstream involves "substance separation" and "product processing". Xiong Lei and his team's efforts in technological R&D

Fan Xiaojun, Professor, Doctoral Advisor, and Director of Department of Business Administration, School of Management, Shanghai University.

lay a solid foundation for business model innovation in the future. But techno-logical innovation usually faces the challenges of high cost, scarce supporting resources, and low market recognition. So technological innovation is likely to fail if not accompanied by corresponding business model innovation. Especially for emerging industries and transformational technologies, the promotion and appli-cation of technologies and products is extremely difficult due to immature tech-nology, high R&D costs, and a lack of supporting facilities. Therefore, effectively reducing cost through business model innovation is an important means for new technologies and new products to enter the market. This applies particularly to emerging industries, because their technological model and business model are still under exploration, and active business model innovation is needed to promote technology application. 3DMed cooperates with hospitals to adopt the "group buying model" for technical services, and signs research-related service contracts at lower prices. Then it cooperates with upstream suppliers with large service con-tracts, which enables technological R&D along with the angel financing obtained by 3DMed.

Though 3DMed faces challenges, including sustainable and high R&D invest-ment, putting a team together after its strategic transformation, and cultivating customers in a changing industry, 3DMed lays a solid foundation for creating a positive enterprise and brand image, with a business philosophy focusing on social marketing, including providing new drug R&D services, building a drug R&D platform, and changing clinical diagnostics models. With this philosophy, as long as 3DMed can better satisfy customers than its competitors by following innovative ideas and adjusting its product portfolio and marketing portfolio strategies based on changes in the market, as well as protecting consumers' interests and enhancing social welfare, 3DMed can definitely achieve long-term and sustainable development.

Case Analysis II

Change Always Works

Lian Minling

3DMed is a typical example of technological team-based entrepreneurship. Its story was just like the first half of a Hollywood blockbuster. Once upon a time, there was a smart young man who dreamed of finding treasure. By chance, he heard of a castle where tremendous riches were hidden, and he immediately decided to give up his cosy and comfy life. He rethought his goals, and set off in search of the treasure. On his way to the castle, he went through an existential crisis, and a few of once-loyal

Lian Minling, EMBA 2014 student at CEIBS and Chairman of Shanghai Longly Venture Capital Co., Ltd.

partners left him. He finally met a powerful nobleman, who gave him a magic sword which enhanced his abilities and enabled him to journey even faster towards the treasure. However, the questions remained in his head: just how far away is the castle? Which road should I take? Is the treasure guarded by a fearsome dragon? Meanwhile, some of his companions continued to abandon the journey due to disagreement, for no one could prove that the castle even existed, nor could anyone say for sure how valuable the treasure might be. What should the young hero do next? For the whole story, please wait for the second half.

In the real life story, the smart young man is Xiong Lei, founder of 3D Med. When he was studying for his Ph.D., he accidentally found the treasure, the personalised anti-cancer drugs that many people were searching for. The investor is the powerful nobleman, and capital is the magic sword. Like most entrepreneurs, Xiong Lei also went through the following phases. The first was survival. In order to survive, he temporarily set aside his goals and started out with small-scale business. He knew that he had to build a financial base before making further growth, so he began with building an RNAi library to sustain his team. However, the spark of his dream kept shining, and whenever the chance arose for him to follow it, he would take up the challenge and hit the road without hesitation. Therefore, after finishing funding, he decided to initiate a strategic transformation of the company, shifting away from the existing profitable business toward precision therapy. He would have great difficulties in carrying out this transformation, facing challenges in R&D and competition with those who had vested interests in the existing medical system. So although his dream was big, the reality he faced was cruel. Failing to understand the transformation, three core members of the founding team chose to quit, which made Xiong Lei's challenge even bigger.

At that time, he faced a critical test, a test for both himself and the whole 3DMed team. First to be tested was Xiong Lei's change management capability, because strategic transformation is a vital issue, and failure to carry it out effectively could cause huge damage. On the one hand, losing founding members must have been a heavy blow to Xiong Lei's company, and on the other hand, being in a highly specialised industry, it was hard for him to find existing qualified talent in the job market. The founding members that left did so partly for their own reasons, but partly due to Xiong Lei. Technical leaders often do not pay enough attention to interpersonal communication. If they cannot form a common goal with their team and make the members believe that the transformation is the right and wise thing to do, this will have a strong negative influence. Most start-ups have the problem that founding members can work together to overcome hardship, but they cannot share happiness—they can work together for a certain period of time, but they often break apart in times of change.

Although Xiong Lei took some measures to fill the personnel shortage arising from the changes in team members, it seems that he paid a relatively high price for it, because the transformation slowed down. This turn of events illustrates well the truth in the saying "more haste, less speed". Indeed, had Xiong Lei paid more attention to the potential resistance, the transformation process might have been less bumpy. Generally speaking, change management comprises three steps: early communication, gradual adjustment and timely reflection.

Firstly, it is recommended that early communication should be carried out progressively, to explain the reasons for the transformation to each organisation and department. This process may include communication within formal organisations, as well as more emotional communication by way of opinion leaders in informal organisations. The objective is to crystallise the reasons for transformation, rebuild the company's vision, and reduce resistance.

Secondly, instead of setting a tight transformation schedule, set monthly adjustment goals to make steady progress while avoiding sudden major changes that may disrupt personnel.

Thirdly, provide a timely summary of successful experience and set role models to inspire other employees. Small achievements made in steps can be more motivating than a single major achievement that takes a long time to accomplish.

In fact, there is a saying in management that the more you fear something, the more likely it is to happen. My understanding of this paradox is that the fear of a negative result can cause people to get nervous and act more passively in the hope of avoiding it. However, this usually makes things worse, and may be more likely to bring about the result that caused the worry in the first place. So if you can confront and tackle upcoming problems immediately, you can avoid the situation described by the saying. The fate of 3D Med is in Xiong Lei's hands, and his moves will have a direct effect on the future development of the company, which has so far been just like the plot of a Hollywood blockbuster. We hope, of course, that Xiong Lei can make the transformation a success, and ensure a happy ending in true Hollywood style.

Case Analysis III

Focus on Entrepreneurship Challenges from Precision Medicine

Sun Zikui

After the completion of the Human Genome Project (HGP) and with the development of high-throughput DNA sequencing technology, more associations between diseases and genes have been discovered, and genetic diagnosis and treatment will therefore undoubtedly create a huge market. According to the World Cancer Report 2014, the number of registered new cancer cases and deaths in China was 3.065 million and 2.205 million respectively in 2012, accounting for around 1/5 of the world's new cancer cases and around 1/4 of cancer deaths globally. The 5-year survival rate of cancer patients in the US was 85%, but that of Chinese

Sun Zikui, EMBA 2015 student at CEIBS, Chairman and General Manager of Shanghai Personal Biotechnology Co., Ltd.

patients was only 25%. Against this background, personalised cancer therapy based on individual genetic testing has become a trend, which means that 3DMed entered the market at the right time.

Most people start a business based on their own specialist area, and so did Dr. Xiong L. In the very beginning, 3DMed aimed to provide services for building an RNAi library, after all, this was Xiong Lei's speciality. However, the market size was small and filled with fierce competition, so it was hard to develop a differentiated competitive edge. In light of this, 3DMed moved on to drug marker screening using NGS. By building tumour cell models, it used genetic sequencing technology to screen targeted drugs, in other words, to find appropriate drug biomarkers to support the R&D of new drugs. This is part of the field of precision medicine. This will be the future trend in disease diagnosis and treatment, so there is a huge market potential for such services. However, the biggest challenge facing precision medicine is the uncertainty of associations between genes and diseases. This is especially true of cancer. Even for the same type of lung cancer patients, their genetic mutation may be quite different, which makes it hard to accurately identify the association between the gene and the corresponding disease. It is therefore necessary to collect a large pool of samples, carry out genome-wide or exome and even transcriptome sequencing on these samples, and use big data analysis to define the type of gene mutation. This means that a huge amount of capital is a must. For a start-up, capital is a major challenge. Collecting samples can also prove difficult.

Besides, I disagree with that view mentioned in the case that "traditional single-gene testing will become a thing of the past", and here are two of my arguments.

Firstly, the total price of a single-gene test could be up to RMB 2000 due to high clinical application fees, but the cost of actually performing the test is no more than RMB 100, which means that there is huge potential for cutting down the price.

Secondly, some diseases are caused by point mutations in single genes. For example, in most cases, hereditary hearing loss is caused by a single gene, so using high-throughput screening technology to test for this kind of disease is unnecessary.[39]

References

1. Jemal A, Bray F, Center MM et al (2011) Global Cancer Statistics. Oncol Radiother (Feb)
2. Chen W. Chinese Cancer Registry Annual Report 2013, Chinese Cancer Registry Center, 14 Apr 2014
3. American Cancer Society (2010) Cancer facts & figures 2010. American Cancer Society, Atlanta
4. DeSantis CE et al (2014) Cancer treatment and survivorship statistics, 2014. CA Cancer J Clin 64:252–271

---

[39]Wang and Wang [37].

5. Berkrot B (2011) Success rates for experimental drugs falls: study. Reuters (14 February)
6. Pammolli F, Magazzini L, Riccaboni M (2011) The productivity crisis in pharmaceutical R&D. Nat Rev Drug Discovery 10
7. Mao K, Chen D. The history of anti-cancer drug R&D. Shanghai Technology Inquiry and Consulting Center, CAS, 22 Feb 2012
8. Lu Y, Chen D, Xiong Y (2012) Analysis of R&D situation of anti-cancer drugs. Chin Bull Life Sci 6
9. Moffat JG, Rudolph J, Bailey D (2014) Phenotypic screening in cancer drug discovery—past, present and future. Nat Rev Drug Discovery 13:588–602
10. Deng Z (2013) Overview of new drug development process. Sci Monthly 2
11. Roche. From Idea to Medicine—Drug Development at Roche. Roche Global, 2 Jan 2013
12. Paul SM et al (2010) How to improve R&D productivity: the pharmaceutical industry's grand challenge. Nat Rev Drug Discovery (March)
13. Wang B. Research on competitive strength of Chinese Biopharmaceutical Industry, 2010 edition published by Shanghai University of Finance & Economics Press
14. Li P. Identifying Future Competitive Trends from Global Anti-Cancer Drug Sales Ranking. Global Oncology Express, April 2014-110
15. Shi J. AstraZeneca Opens an Innovation Centre with Huge Investment. Xinmin Evening News, 23 Mar 2007
16. Yu D. Roche Launches China's First Comprehensive Clinical Trial Centre. The Economic Observer, 31 Oct 2007
17. Zhu Q (2009) Lilly's Unique R&D System. CEIBS Bus Rev 7
18. Yang J. Johnson & Johnson Establishes an R&D Centre in Shanghai as Its Asia Pacific R&D Headquarters. Science and Technology Daily, 23 Apr 2009
19. Xu H. Novartis Invests USD 1 Billion to establish China's largest drug R&D centre. Economic Daily, 24 Nov 2009
20. Liu C. Ding Jian: Chinese pharmaceutical companies lack anti-cancer drug R&D capability. China Science Daily, 12 June 2012
21. Wang J. Branded drugs perform better than generic drugs. Wen Wei Po, 30 Apr 2014
22. Naisbitt J (2009) China's megatrends (trans: Wei P). China Industry & Commerce Associated Press in 2009
23. Chen S (2013) Somatic mutation and novel strategies for personalized targeting therapy. J Zhejiang Univ 42(1)
24. Zhou H, Liu J (2011) Gene targeting tailored therapy for the anti-tumor drugs. Anti-Tumor Pharm 1
25. Wang K, Liu B (2012) Truly personalized medicine to combat cancer. Medicine & Philosophy (Clinical Decision Making Forum Edition) 6
26. Luo B (2003) Application progress of pharmacogenomics. Chin J Pract Med 19
27. Sun H, Ouyang W, Sun J (2004) Trend in research of pharmacogenomics in foreign countries. Chin J New Drugs 3
28. Calvo KR, Liotta LA, Petricoin EF (2005) Clinical proteomics: from biomarker discovery and cell signaling profiles to individualized personal therapy. Biosci Rep 25(1–2)
29. Sherman RE et al (2013) Expediting drug development—the FDA's new 'breakthrough therapy' designation. N Engl J Med 369:1877–1880 (14 Nov)
30. Subramanian R, Sheppard T, Rubin B, Kramer C (2013) FDA's new breakthrough therapy designation: what does it mean for pricing and market access? OBR Green, Sept 2013, vol 7 (8)
31. Gangwar SK, Worabo B (2011) Amazing facts about human DNA and genome. Society for Science and Nature, 20 Sept 2011
32. McLeod HL, Evans WE (2001) Pharmacogenomics: unlocking the human genome for better drug therapy. Annu Rev Pharmacol Toxicol 41:101–121
33. Han C, Yang SC (2005) High throughput screening assay and application. Biotechnol Bull 2
34. Liu L (2012) Illumina: the 'Apple' that disrupts the gene sequencing sector (2 July 2012), Instrument Information Website, http://www.instrument.com.cn/news/20120702/079802.shtml. Last accessed 3 Nov 2015

35. Herper M (2011) The No. 1 Winner in DNA Sequencing: Illumina (5 Jan) (trans: Yu B). Forbes China, http://www.forbeschina.com/review/201101/0006529.shtml. Last accessed 3 Nov 2015
36. Sharma SV, Haber DA, Settleman J (2010) Cell line-based platforms to evaluate the therapeutic efficacy of candidate anticancer agents. Nat Rev Cancer 10:241–253
37. Wang H, Wang Q (2014) Application of and prospect for targeted capture with the first generation of sequencing technology in the study of hereditary hearing loss. J Audiol Speech Pathol 5

Case II: Micro Platform, Major Innovation—WeChat-Based Ecosystem of Innovation

The end of 2010 welcomed the debut of several mobile instant messaging tools. Interactive Technology launched "iGexin" on November 7th, 2010. Tencent set up the project of "WeChat"[1] on November 20th. Xiaomi Technology introduced "Miliao" on December 10th.

In the era of mobile Internet, it is necessary to "choose the right innovation at the right time" to make a success. During the past four years, by virtue of Tencent's rich social network resources, WeChat carried out targeted innovation and rapid iteration, growing from an average messaging app to an omnipotent platform serving basic needs of everyday life. As a phenomenal product generating profound social and economic effects, WeChat has nurtured the largest ecology in the mobile Internet.

WeChat is now unlocking the potential of its official account platform in marketing and e-commerce, attracting numerous developers, traditional companies and startups to come and gain a foothold. WeChat is laying stepping stones for the grand vision of "mass entrepreneurship and innovation".

Precision Innovation and Rapid Iteration

Founded in 1998, Tencent Holdings Ltd. has long been dedicated to the development of social network, and introduced the first instant messenger in 1999. The company was listed in Hong Kong in 2004, when the majority of its profits derived from the value-added services of China Mobile and Telecom. It earned the revenue

---

This case was prepared by Professor Zhu Xiaoming of CEIBS, part-time case research fellow Li Yang, research assistant Ren Yifan, and part-time research fellow Song Yanbo based on public materials. It is intended as a basis for class discussion rather than an illustration of an either effective or ineffective handling of a management situation.

---

[1]Zhang [1].

© Springer Nature Singapore Pte Ltd and Shanghai Jiao Tong University Press 2018
X. Zhu, *China's Technology Innovators*, Management for Professionals,
DOI 10.1007/978-981-10-5388-7_2

of RMB 19.6 billion in 2010, most of which came from Internet value-added services (games). The instant messenger QQ run on PCs used to be the core product of Tencent.

At the end of 2010, the organic integration of Internet, mobile Internet and smart terminals intensified the competition in mobile instant messaging market. By that time, mobile Internet users had accounted for 66.2% of the total number of netizens in China.[2] The penetration and adoption rates of smart terminals continued to climb in China, and the habits of Internet users began to change, ushering in the new era of mobile Internet.

Ma Huateng, Chairman and CEO of Tencent, voiced his opinion of this new trend:

> To gain an upper hand in the era of mobile Internet, we can't simply rely on any single product of any department. All departments should work to ensure that all their products can be run on both PCs and mobile terminals, so as to fully embrace mobile Internet… However strong you are, the overwhelming wave of mobile Internet can easily capsize your boat of business, should you have any negligence. Therefore, we should always hold industry evolution in awe and commit ourselves to providing better services.[3]

Tencent is like a super incubator that has given birth to over 1700 products.[4] When WeChat was set up as a new project, the QQ team was developing another mobile instant messenger QChat. Eventually, the development of WeChat left QChat in the dust and the QChat project was called off.[5]

Up to now, WeChat has eclipsed QQ and is valued at USD 64 billion.[6] WeChat has become an integral part of Chinese people's daily life. According to a survey in 2014, 69% of the interviewees used PC chat tools to communicate, while 80% used WeChat for communication.[7]

The story of WeChat started from a simple idea. On November 19th, 2010, Zhang Xiaolong, General Manager of the R&D Dept. of Tencent Guangzhou, wrote on Weibo:

> My only expectation for iPhone5 is that it doesn't support phone call just like iPad (3G). In this way, I can save on phone bills while you can still text me through Kik, call me through Google Voice, and video chat with me through Facetime.[8]

The WeChat project was kicked off on the next day.

Mr. Zhang is a senior product manager who developed Foxmail in 1997 and then took charge of QQMail after Foxmail was acquired by Tencent. Since the launch of the WeChat project, rapid iteration has brought along many innovative features. By

---

[2]CNNIC Report Analysis [2].

[3]Ma [3].

[4]Nan [4].

[5]Ma [5].

[6]WeChat Valued at USD 64 billion, Three Times the Value of WhatsApp, zol.com, March 13, 2014.

[7]Zhang [6].

[8]Zhang [1].

July 2012, Mr. Zhang's original ideas (functions of text, phone call and video chat) for the WeChat had been completely brought to life through the release of its version 4.2.

WeChat's major versions and their features are as follows:

Version 1.0: The slogan of "free SMS with photos" failed to attract customers due to the inexpensive monthly plans offered by mobile operators.

Version 1.2: The feature of sharing pictures was introduced.

Version 2.0: The feature of voice chat found favor with customers. WeChat-related posts popped up on Sina Weibo on a per-minute basis.

Version 2.5: The feature of "People Nearby" was introduced, enabling social interaction among strangers through voice service and LBS (Location Based Service).

Version 3.0: The feature of "Shake" was introduced. The number of users grew exponentially, establishing WeChat as a market leader.

Version 3.5: The feature of "Scan QR Code" was introduced.

Version 4.0: The feature of "Moments" enabled users to share texts, pictures, music and videos in an intimate circle. WeChat opened APIs in order to build a mobile social platform.

Version 4.5: Features of voice/video call, Web WeChat, following/subscribing official accounts were introduced.

Version 5.0: Features of WeChat payment, Sticker Gallery and Games were introduced to tap into business opportunities. The upgraded feature of Scan QR Code could provide useful information after scanning barcodes, English texts, book covers, and streetscape. The game "Plane Fight" soon became a hit upon its release.

...

WeChat released 45 versions for different terminals in 2011, updating almost once a week. It evolved from a provider of "free SMS with photos" to "the most favored mobile instant messenger". Although its speed of updating slowed down after 2012, micro innovations have paved the way for structural platform innovation, which triggered powerful disruptive effects.

WeChat grew into one of the top social platforms in China in four years, tapping into the market outside the sphere of Tencent and establishing itself as a mobile Internet platform. The growth of WeChat outran global social platforms in 2014. From June 2013 to June 2014, the adoption rate of WeChat (overseas version) was up by 26%, leaving other social platforms far behind. In comparison, Instagram saw an increase of 18% in the adoption rate, Facebook and Twitter registered zero growth, while YouTube, Pinterest and Tumblr suffered negative growth[9] (see Fig. 1).

---

[9]Zhang [7].

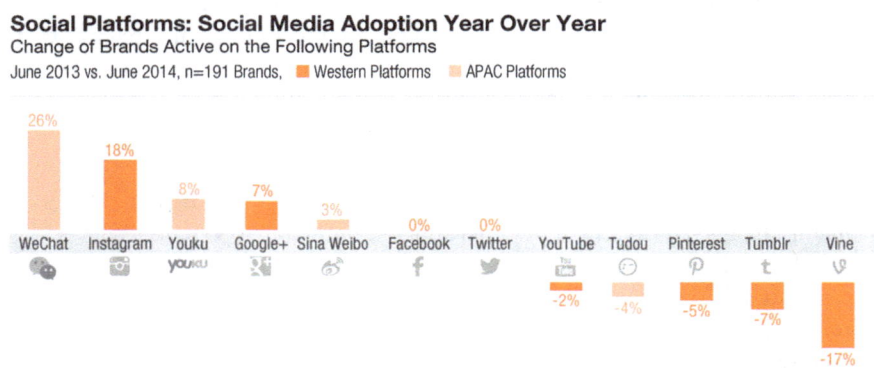

Fig. 1 Changes in Social Media Adoption Rates (June 2013 to June 2014). *Source* L2 Intelligence Report, "Social Platforms: L2 Assesses The Social Investment And Performance of 382 Brands Across 17 Platforms", http://www.l2inc.com/social-platforms/

Methodology of "Micro Innovation"

A small product can have a large market. With micro innovations cropping up one after another, WeChat serves as a typical example of disruptive innovation within a big company.

WeChat has integrated social interaction, e-commerce, payment, and O2O features into one platform. It offers various functions such as instant messaging, Moments, Contacts Sync, QQ Mail alert, Weibo message alert, Message in a Bottle, People Nearby, Voice Notes, Shake, Broadcast Messages, access to Weibo, mobile data check, game center, access to JD.com, official accounts, WeChat payment, intelligent hardware interface, and JS SDK.

Many of these features are not original. The WeChat team has applied others' original ideas to its own product and strived to provide the best user experience. As Zhang Xiaolong explained:

> After the release of WeChat 4.0, many industry critics accused us of plagiarizing Instagram or Path. However, few discovered the exquisite beauty of the feature Moments. They didn't realize what a risky move it was to provide social networking service on the basis of QQ relation chain, and how we tried to avoid the potential harm to user experience by making regional improvement, let alone the possible structural change brought by the third-party content introduced through open interfaces. When they attempt to cover up their mediocre performance and refuse innovative ideas by accusing us of plagiarism, they are falling further behind.[10]

Examples of micro innovation abound during the iterative development. For instance, with the improved version of Voice Chat, you can hear voice messages

---

[10]Sun [8].

through the receiver when the proximity censor is activated and the loudspeaker is only turned on when your ear is away from the phone. This is a considerate way to keep user privacy from being broadcast in the public.

The WeChat team has formed its own methodology to achieve precision innovation and rapid iteration.

A Decentralized Organizational Structure

Innovative ideas often stem from the lower level of an organization. In many companies, it's the profit-making department that has the biggest say, and this is of no help to their long-term development. In Tencent, new departments are independent of those cash-cow departments, namely, wireless business and interactive entertainment systems. The WeChat team has little to do with the core businesses of the company. Most important, WeChat, as a project of disruptive innovation, was not screened out or stashed away by the frustrating evaluation process.

Small Teams Led by Project Managers

Different from the standardized manufacturing process, micro innovations cannot be carried out on the assembly line. Small capable teams and breakthroughs of the conventional process are the defining features of micro innovations. Mr. Zhang persists in the mode of small teams by fully empowering each team and streamlining the cooperation process, so as to motivate team members.

Innovate and Rapidly Iterate by Keeping Abreast of the Times

The technologies used by WeChat are simple but not original. WeChat bases its features on structural functions built in smart phones. For example, Voice Chat relies on microphone and loudspeaker, People Nearby on GPS module, and Shake on gravity sensor module. These innovations have everything to do with the recombination of underlying technologies and function modules for the sake of better user experience. While upholding the principle of "Simplicity is Beauty", the WeChat team attends to user demands and works on details to improve the product. These practices seem easy to copy, but the fact is that none of the similar applications can keep up with the rapid iteration of WeChat.

Less Is More and Reverse Iteration

WeChat has never blindly added the features recommended by users. Instead, it has cut certain features to make the product more user-friendly. For instance, iPhone's iMessage can send read receipts to users. The WeChat team didn't follow suit because they didn't want to disturb users with a feature that may force the receiver to reply immediately. They put much emphasis on reverse iteration, as they are aware that a product is not welcomed for its numerous features, but for its convenience to use.

Dynamic Management and Targets for Different Stages

Good ideas sprout up all the time and users tend to abandon old products for new ones. With the ever-changing market, advantages in resources today may soon be rendered obsolete. Moreover, the high turnover of the WeChat team members necessitates dynamic management and formulation of targets for different stages.

Open Interfaces and Set up the Platform

Tencent used to copy competitors' products and grabbed market shares from many SMEs whose products were eventually forced out of the market. In recent years, Tencent has adjusted its strategy and gradually set up an open platform on http:// open.qq.com/ and released Internet Open Platform White Paper, offering access to third-party applications. WeChat is now dedicated to establishing a platform by opening API and introducing commercial opportunities.

Innovation of Platform Service

WeChat has become the new soul of Tencent. As the part and parcel of Tencent's strategic layout, it explored international business, stood against the expansion of Baidu and Alibaba Group, and acquired shares in Dianping and JD.com. On May 6th, 2014, Tencent announced the establishment of WeChat Business Group and appointed Zhang Xiaolong as its president, signaling WeChat's transition from a mobile social networking product to a business system of strategic importance.

Openness serves as the prerequisite for a mobile Internet platform like WeChat. Ma Huateng, Chairman and CEO of Tencent, explained why WeChat was positioned as a platform.

We put more focus on the economic system (than the product alone), so we want to build a platform and invite partners to join and develop it... Sustained growth

Table 1 Nine open APIs offered by WeChat

Open Access	Description
Speech Recognition	Recognize and translate users' voice messages into text
Customer Service Interface	Official accounts can reply users within 12 h after receiving users' messages
OAuth 2.0 Authorization	Official accounts can request user authorization
Parametric QR Code	Official accounts can get QR codes with different parameters and analyze their effects through parameters after users scan codes to follow accounts
Users' Location	Official accounts can access the location of users when they open accounts (upon user approval)
Users' Basic Information	Official accounts can obtain users' basic information including profile photo, name, gender and region through encrypted OpenID
Followers' OpenID	Official accounts can obtain the OpenID of all its followers
User Group	Official accounts can group users, create and edit groups in the backstage
Upload or Download Multi-Media Files	Official accounts can upload and download multi-media files through the WeChat server when needed

can only be achieved when you are able to bring new business opportunities onto the platform and bring mutual benefits to your partners and yourself.[11]

Take the nine open APIs offered for free for service accounts for example. Companies like China Southern Airlines, China Merchants Bank, Guangdong Unicom have used these open interfaces to provide customized service and even "intelligent customer service". The official account of "Haier Intelligent Air Conditioner" offers the feature of speech recognition, through which WeChat receives instructions from the user and then sets the temperature, fan speed and mode of the air conditioner. The official account of iCNTV also adopts this feature and allows users to switch TV channels through speech recognition, which is more convenient than the remote control (see Table 1).

WeChat also optimized the way of online communication. The WeChat team refined the model of Path to provide better user experience through the feature of Moments. User relations fall into several categories and only friends on the strong relation chain can see and comment each other's posts. Contents in different relation chains are separated from one another under precise control.

These refined innovative features dealt a blow to such competitors as Miliao and Sina Weibo. The user activity rate of Sina Weibo plummeted by at least 30% in 2012, and meanwhile, more than 60% of the 300 million WeChat users were active in the Moments.[12]

---

[11]Ma [3].
[12]Sun [8].

WeChat's innovation is carried out in four categories:

Service Integration: O2O service innovation in the business sector;
Ecosystem: Opportunities for entrepreneurship and innovation born from the official account platform;
Rapid Iteration: Product innovation such as Shake and Moments;
Progressive Innovation: Innovation with underlying technologies such as Intelligent Recognition and data application.

Based on the social network, WeChat has set up a platform providing services for everyday life, which is closely related with traditional industries. Its official account platform has created a wealth of opportunities for entrepreneurship and innovation. And it also supports O2O service innovation to promote businesses.

Similar to Weibo, WeChat plays a role of medium. According to the survey, more than 60% of users open WeChat over 10 times each day. User loyalty has made WeChat the main channel for users to obtain information. 73.4% of users follow official accounts while 40% get information from official accounts, WeChat groups and Moments.[13] The official account platform has drawn the authors who had faded away with the era of blog back to contribute original articles.

From July 2013 to June 2014, the volume of information consumption driven by WeChat stood at RMB 95.2 billion, equivalent to 4.24% of the national total in 2013. Apart from data consumption, which is the biggest share of WeChat's contribution to information consumption, games and official accounts on WeChat have also seen the rise in consumption.[14]

WeChat has taken gradual steps to cash in on its features. Game distribution, IP traffic export, financial products, paid stickers, and ads on mobile terminals were introduced to make profits. WeChat 5.0 classified official accounts into subscription accounts and enterprise accounts. More and more businesses have opened up official accounts to attract users and provide rich user experience. Since WeChat can help companies establish proprietary communication channels, it's reasonable to collect service fees (see Fig. 2).

WeChat has carried out service innovation in such areas as multi-customer-service system, WeChat stores, WeChat ads, Wechat logon, intelligent platform and hardware.

As the fast track for small businesses to open up stores, WeChat stores provide native support for store keepers. Presently, 95% of the businesses that have enabled WeChat payment have set up WeChat stores. WeChat has cooperated with the advertising platform Guang Dian Tong to sell WeChat ads, attracting over 10,000 advertisers. Targeted promotion for advertisers is made possible via WeChat's big data.

---

[13]"Report on the Social and Economic Impacts of WeChat", 199it.com, January 27, 2015.
[14]"Report on the Social and Economic Impacts of WeChat", 199it.com, January 27, 2015.

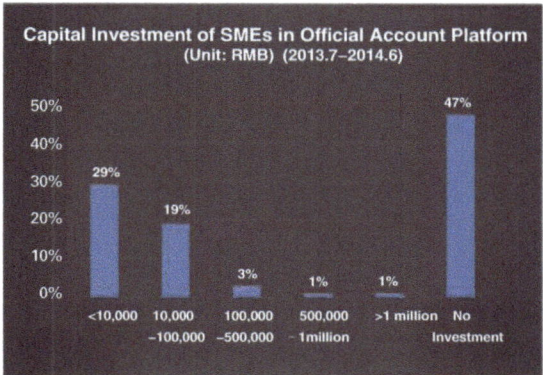

53%

Among the companies and organizations that have opened official accounts, 53% have invested in the account operation or the development of advanced features.

Fig. 2 The Impact of WeChat Official Account Platform. *Source* "Report on the Social and Economic Impacts of WeChat", 199it.com, January 27th, 2015

An O2O Closed Loop Formed by WeChat Payment

The survey shows that 20% of the brands that choose to cooperate with WeChat are attracted by the brand loyalty this platform can help build.

Different from their digital marketing approaches on other social media such as Facebook and YouTube, luxury brands put much focus on how to make full use of the highly interactive feature of WeChat. The setting of two-way following and privacy of individual information have increased the reach and effectiveness of promotion on WeChat. Louis Vuitton opened up an official account to provide one-on-one customer service. Cartier official account can offer users immediate information about the nearest store, its map, contact and navigation by means of Location-Based Service. Consumers can also use the function of "translation" to check product information and communicate with shop assistants while shopping overseas.[15]

The steady commercialization of WeChat has something to do with the growth of its infrastructure, WeChat payment.

Since the launch of WeChat 5.0, the team has been committed to the strategy of building a platform. WeChat payment allows users to pay bills by scanning QR codes, and it also can be used for in-app purchase and payment in official accounts. WeChat has introduced third party operators such as Dangdang, Youku, Dianping, and JD.com, and enabled users to top up phones, buy movie tickets, lottery, and coffee through the platform. On November 28th, 2013, WeChat jointed hands with Xiaomi in launching a flash sale, during which 150,000 Xiaomi phones were snapped up in less than 10 min.[16]

---

[15]Zhang [7].

[16]Flash Sale of Xiaomi on WeChat [9].

Social Platforms: WeChat Post Engagement by Category

August 2014, n=170 Accounts

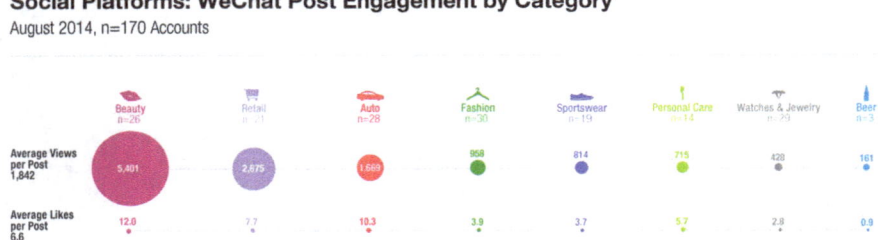

Fig. 3 WeChat Post Engagement by Category. *Source* L2 Inteligence Report, "Social Platforms: L2 Assesses The Social Investment And Performance of 382 Brands Across 17 Platforms", http://www.l2inc.com/social-platforms/

All the functional features of WeChat platform, namely, WeChat payment, LBS, membership service, sample display, voice chat, and live chat, can generate profits. For instance, thanks to product display and membership service offered through the WeChat platform, beauty brands have earned the most views and "likes" per post (see Fig. 3).

WeChat payment is the corner stone for its path of commercialization. WeChat has focused on "one-click payment" and formed an O2O closed loop, posing a direct challenge to Alipay. In 2013, it took three months for WeChat payment to obtain 10 million users from scratch. Today, it's still growing by attracting over 100,000 new users per day.[17]

The growing influence of WeChat payment was evidenced by Taobao's attempt to block it. Since many Taobao stores have official accounts on WeChat, as long as they link bank cards to WeChat, they can easily execute transaction and payment by scanning QR codes on WeChat without paying extra fees. Moreover, Tecent's E-commerce Department provided tens of thousands of offline stores with QR codes for WeChat payment, forcing Alibaba to fight back by blocking visits from WeChat under the pretext of security concern.

WeChat Red Packet was introduced with much fanfare as a way to boost WeChat payment. Red packets that are sent to each other based on users' social relation chain encourage users to link bank cards to WeChat payment, thus making this feature a secret weapon in the battle of mobile payment. The popularity of red packets can tell a lot about the growth prospect and business potential of WeChat payment.

Tencent has set up a team composed of staff from nearly 20 departments to ensure smooth operation of red packets. Another feature "Shake" is also pivoting to commerce. Apart from "shaking" to find people, users can now opt for "Nearby" to shake for red packets, coupons offered by nearby stores and their indoor navigation.

WeChat jointed hands with businesses to deliver random red packets to users through "Shake" during festivals. On the New Year's Eve of 2014, WeChat users

---

[17]War of Payment [10].

shook 4.82 million times to receive more than 20 million red packets, which was 9412 packets per minute. In the Spring Festival of 2015, "Shaking for Red Packets" turned into a national carnival, which saw 7.2 billion times of shakes (810 million times per minute during the peak) and 120 million red packets given away.[18]

Global Positioning and Globalization

Tecent's WeChat (overseas version), America's WhatsApp, and Japan's Line are recognized as the three giants in the field of mobile instant messaging. These products share the features of free messages, voice chat, social networking based on LBS, and contacts sync, but differ in product ideas, business models and cultural backgrounds (see Table 2).

WhatsApp, an application famed for clean design and dedicated to communications service, has positioned itself as a SMS alternative since its inception. With the slogan of "No Ads, No Games, No Gimmicks", WhatsApp promises never to introduce ads or fashionable features like emoji, but to serve group or private chats only and keep users always online.

Users can enjoy the first year of WhatsApp service for free, and pay USD 0.99 per year after that. In Taiwan, however, the free app Line posed a challenge to WhatsApp. The co-founder of WhatsApp Brian said, "We changed the business model in Taiwan by making it free, but we've already lost the chance there."[19]

Facebook, Tencent's major rival in global instant messaging market, announced on February 19th, 2014 that it would acquire WhatsApp for USD 19 billion. WhatsApp suffered a deficit of USD 230 million over the first half of 2014 and slowed down its process of commercialization after acquisition.

Compared with WhatsApp, Line is more like WeChat, boasting rich functions such as free calls, free messages, status of always online, stickers and emoji, change of background pictures and group chat, with interesting stickers as its most prominent feature. By October 2014, Line had attracted 560 million users worldwide with 170 million active user accounts. Apart from Japan, Line has made its way into markets of Thailand, Indonesia, Spain and Taiwan.

Line has also produced dozens of derivative products which run independently, including postcard, camera and picture editing, painting, location sharing, interest group, three social games, anti-virus protection, private social network, weather forecast, manga reading, etc. According to statistics, Line users, on average, send a sticker every six sentences in chat sessions. Line pockets over USD 10 million per month by selling stickers. Opening official accounts for companies and stars and

---

[18]810 Million "Shakes" per Minute for Red Packets During the Spring Festival Gala, Beijing Times, February 19, 2015.

[19]Nan [11].

Table 2 Comparison of Mobile Instant Messaging Products

App	Country	Number of Users	Active Users	Services	Business Model	Cultural Background
WeChat	China	About 700 million	438 million	Integration of social networking, e-commerce, payment, and O2O features	Ads and game distribution	Profound Chinese Culture
WhatsApp	US	Nearly 1 billion	700 million	Confine to group or private chat, featuring voice chat, text, photo, video, and location sharing	Annual fees of USD 0.99	American Puritanism
Line	Japan	560 million	170 million	Develop Line as the core product while introducing derivative products	Games, stickers and ads	Manga Culture

All data came from the latest official data release of these apps. The user numbers of WeChat and WhatsApp were the estimation by referring to survey agencies' data

Source Nan Qidao, "Three Giants of Social Networking: What are the Differences Among WeChat, WhatsApp and Line?", Tech 163, January 29th, 2015

selling ads are also important sources of its revenue. As its major cash cow, the game business contributes more than half of Line's total revenue.

WeChat functions more like a platform than WhatsApp and Line.

It's easier for mobile communications service to cross borders and reach global users than search engines and e-commerce. WeChat has been translated into over 20 languages and attracted more than 40 million overseas users. It topped the rankings of App Store's most popular social networking apps in 15 international markets in 2012. Tencent is vigorously expanding its international business by tapping into Southeast Asia, Latin America, South Africa, Europe and the US. To promote the product in Europe and the US, WeChat promised to reward the user with a 25-dollar gift card if he or she could invite five friends to log in WeChat.[20]

The advertising revenue of Facebook was over USD 5.5 billion in 2014, more than 66% of which came from its mobile advertising. According to Tencent's financial statement, its advertising revenue in Q3 2014 was RMB 2.44 billion, only 12% of the total revenue. These figures show the great potential of mobile advertising.

Mass Innovation Inspired by the Official Account Platform

"WeChat is a life style," claimed its official website.

---

[20]Nan [11].

The Internet has restructured the life style of Chinese consumers. For instance, 60% of consumers went to theaters for movies a decade ago. Fast forward to 2014, 69% people watched movies online or after downloading and 49% used mobile terminals to enjoy movies. The Spring Festival Gala, once a must-see TV show on the eve of the Lunar New Year, is now losing its attraction and even reduced to the background music during the festival, which couldn't have been imagined in the past.

What's also beyond expectation is the innovation capacity of Chinese enterprises. Chinese Internet products used to be considered as knock-offs of Silicon Valleys' innovative products. Tencent's QQ, for example, learned a great deal from the pioneer product ICQ. However, Chinese companies often come up with more interactive and interesting innovative services to knock over the "masters" at Silicon Valley.

Tencent reigns in the Chinese market, functioning like an empire composed of AOL, Facebook, Skype, Yahoo, Gmail, Norton and Twitter. WeChat offers more powerful features than WhatsApp in the field of mobile communications. Compared with Microsoft's Skype, users of WeChat can choose whether to reply or when to reply a message. WeChat users can also communicate with strangers through "People Nearby" and share photos with friends through "Moments", a feature like Instagram.

WeChat is evolving into an ecosystem of mobile Internet. Zhang Xiaolong described his vision of this ecosystem:

We hope to nurture a forest, an environment where animals and plants can grow freely… We hope this official account platform is a dynamic system. We don't think a system with rigid fixed rules is a good one. A dynamic system is more likely to achieve dynamic stability. Therefore, instead of doing everything by our own, we invite third parties to build this system with us. WeChat official account platform will see constant changes because this system is always improving itself. It's these changes that help our system obtain dynamic stability… All our concerns are based on one prerequisite, user value first.[21]

Varied features and service innovation of WeChat have brought about abundant opportunities for entrepreneurship and innovation. Consumers can do shopping in different scenarios such as booking hotels and plane tickets, and buying financial products… Each scenario is associated with a traditional or innovative industry. WeChat is like the West during the time of American Frontier, where chances for gold rush abound (see Fig. 4: Entrepreneurial Opportunities Brought by WeChat).

Why this platform can encourage innovation? This should be attributed to Zhang Xiaolong's ideas of "decentralization and disintermediation". WeChat is delegating the power of connecting customers to companies, instead of having everything to itself and charging others for using the platform. Companies need to take the initiative in establishing new models of connecting customers. Having nothing to

---

[21]Zhang [12].

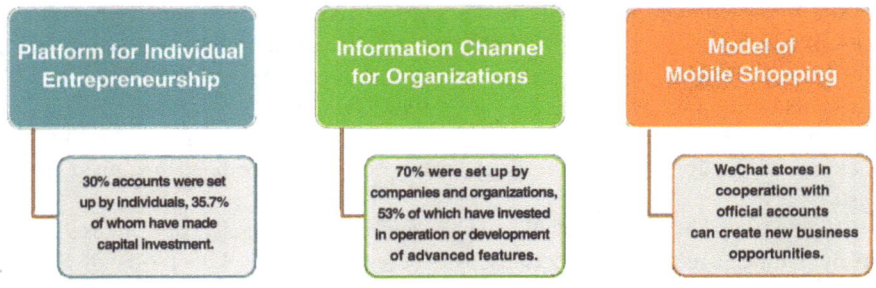

Fig. 4 Entrepreneurial Opportunities Brought by WeChat. *Source* "Report on the Social and Economic Impacts of WeChat", 199it.com, January 27th, 2015

do with scheme or calculation, this design speaks of profound expectation for this fledging platform.

Mr. Zhang's dream has just taken off. Being a platform for entrepreneurship and innovation, WeChat has generated social effects unparalleled by any other similar product in China.

According to statistics, from July 2013 to June 2014, WeChat provided 10.07 million jobs, among which the official account platform contributed 9.78 million jobs and the application platform 300,000 jobs; 1.92 million people were directly employed and 8.15 million were indirectly employed (see Fig. 5). Apart from Taobao.com, WeChat has become another major stage for small startups and entrepreneurs.

The trend of mobile Internet is reaching out to every nook and corner of China's social life. In the fast-changing business world that's always reforming itself with new ideas, more efficient and effective models are successively being proposed and practiced. WeChat is not doubt the masterpiece born from this trend of innovation.

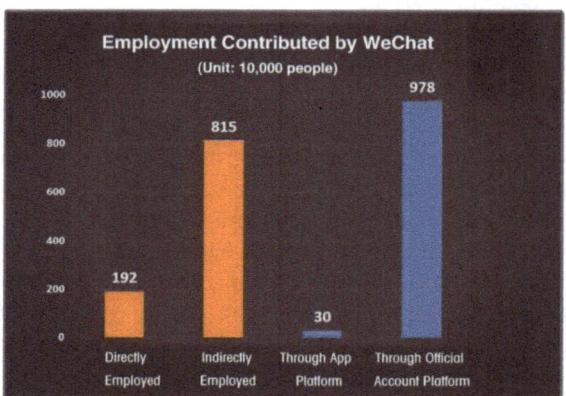

Fig. 5 WeChat's Impact on Employment. *Source* "Report on the Social and Economic Impacts of WeChat", 199it.com, January 27th, 2015

The relation chain and ecosystem of WeChat have aroused the passion for exploration and creation among the general public.

Thanks to WeChat, we are a step closer to the day of "mass innovation".

Case Analysis I

A Successful Example in the Prosumer Age

Ye Weiling

There is a slogan on the official WeChat website which reads "WeChat is a lifestyle". If you consider this slogan to be dull and not very creative, you may not fully understand the notion of "lifestyle". In short, a lifestyle refers to a person's behavioural pattern, which from a business perspective, determines how they allocate time and money. The success of WeChat and WeChat-inspired technologies, as described in the case study, shows how companies can succeed in the Internet age by revising their understanding of consumers.

When the Industrial Revolution began over 200 years ago, as machines began to replace manpower, producers and consumers were divided into two separate categories with distinct roles and functions. However, this clear boundary has become blurred over time. In 1980, the futurist Alvin Toffler created the word "prosumer" in his book *The Third Wave*, to describe the merging of producers and consumers.[22] Over the next 20 years, this concept was not a hot topic for research or practical operations.

Yet as we entered the 21st century with ever-expanding Internet use, Alvin and Heidi Toffler relaunched a discussion on this topic, with a specific section focusing on the "coming prosumer explosion" in their new book.[23] Then over the last decade, the concept went viral. Marketing experts went beyond the notion of "consumer", developing new concepts such as "post-consumer", "value co-creator", "working consumer", "playbor", "digital labor", and "cosumactor" based on Actor Network Theory (ANT), all reflecting the deep insight of theorists into practice and experience.

In fact, consumers have been engaged in production (supply) activities for a long time. When researchers contemporary with the Tofflers studied self-service, they were surprised to discover that rather than resenting the additional labour, consumers actually found it liberating, in particular many "poor" consumers who believed self-service would protect them from the disapproval of store owners or

Ye Weiling, Associate Professor, School of International Business Administration, Shanghai University of Finance and Economics.

[22]Toffler [13].
[23]Toffler and Toffler [14].

staff. So self-service could make store owners and customers more independent and even promote harmony in the community. For the same reason, consumers were also happy to use self-service equipment, which was then further developed and improved. The most typical example is supermarket trolleys.

Nowadays, self-service shopping with trolleys is unremarkable. But store owners in the 1940s didn't like self-service equipment, as their understanding of stores and shoppers was very different from today. Although the promoters said this equipment would save labour costs, the owners still considered service as an investment in customer loyalty, rather than a cost. They were so attached to helping customers choose products, and controlling the supply side, that they considered service levels to be their unique selling point. But in fact there were more and more customers willing to use shopping trolleys, who enjoyed having an open and free shopping environment. They were happy to browse and select products themselves, rather than relying on the 'controllers'.

Using shopping trolleys in grocery stores was considered to be revolutionary. No campaigns were held to introduce them yet consumers were willing to take on some of the labour involved in the shopping process and sales volumes even increased. Then as trolleys were used by more and more consumers, they were continuously improved, and a range of new self-service equipment was developed, such as trolleys with child seats and devices for collecting shopping data. This meant retailers could offer a more diverse shopping experience and acquire valuable information to improve their management.

These days, shopping trolleys have been replaced by mobile phones and websites. The growing number of users has led to the continuous improvement of self-service equipment; and users are willing to take more responsibility on the supply side without feeling like they are working. This unconscious joint value creation ensures more revenue for Internet companies. Now prosumers use various platforms such as Google, Wikipedia, Baidu, Sina, Facebook, Taobao, Amazon, Twitter and WeChat, as well as food delivery, taxi booking and radio applications. Without their contribution, such products would not even be possible.

Although many traditional companies are still reluctant to give control over supply to consumers, just like the store owners who rejected shopping trolleys, as computers and the Internet become more and more popular, the prosumer age seems to be quite unstoppable, just like the Industrial Revolution.[24]

As well as new-generation prosumers, older consumers also want to get in on the act. Indeed, some of them are rewarded for their contribution to production, such as those who work on the Amazon Mechanical Turk (MTurk), a crowdsourcing Internet marketplace. However, even more people are acting as both consumers and producers without receiving any payment, they simply enjoy it. Prosumer characteristics can be summed up as follows: (1) they are not aware of (or don't care about) their participation in supplying products or services; (2) when buying products or services, they are also creating value through labour; (3) they have a

---

[24]Ritzer [15].

conscious positive attitude rather than deliberate resistance towards the whole process of prosumption, for example they enjoy being involved in production.[25]

There is no doubt that the success of WeChat, as well as various innovations by WeChat users, is based on these conscious positive attitudes. However you describe WeChat users, given their wide range of innovations, considering consumers to be independent from producers has become outdated. A traditional mindset tends to put producers and consumers into opposing and separate positions. Whether for practical operations or theoretical research, this mindset contributes nothing to the debate.[26]

In short, WeChat-enabled prosumption can be considered to be the best way of combining production and consumption. During this process, prosumers not only have a certain control over production, they also enjoy the pleasure of consumption. In addition, they may feel less manipulated and can avoid the effects of advertising.[27] Whether online or offline, prosumers get both physical and spiritual benefits from the process of prosumption. For companies that promote the concept of prosumption (such as WeChat and various innovators on the platform), they can benefit by selling relevant products or offering abundant and accurate data.

Case Analysis II

The Vitality of Micro Innovation by WeChat

Xu Qiang

After its launch in 2010, WeChat become the leading Chinese mobile Internet platform in less than four years. This great success could not have been achieved without Tencent's corporate culture that encourages internal innovation. Although Tencent already had QQ, a successful interactive product, they showed great daring and vision when they followed the mobile Internet trend and developed WeChat. The WeChat development team was led by Allen Zhang in the Tencent R&D Centre in Guangzhou, rather than the QQ team, a further indication of the open environment for innovation projects at Tencent. The company spares no effort in supporting each innovation team and project and has been a super incubator for about 1700 products, making Tencent into an undisputed Internet business empire. This shows that internal innovation is of great importance to company development

Xu Qiang, 2013 EMBA student of CEIBS, Chair of Shanghai Qiao & Yun Construction and Engineering Co., Ltd.

[25]Cochoy [16].
[26]Ritzer and Jurgenson [17].
[27]Chia [18].

and survival, in particular Internet companies who have to achieve innovation against the clock.

WeChat is typical in the current "Internet Plus" age of mass innovation. At first, WeChat built their success on services and user experience, which were valued by the product team for achieving "micro innovation" throughout the entire R&D process. Then their open and inclusive attitude towards development allowed them to turn their simple mobile messaging app into a platform product that could support the whole mobile Internet ecosystem.

The concept of "micro innovation" was first introduced by Zhou Hongyi, Chairman of Qihoo 360, for Internet product innovation. As new technologies continued to emerge and mobile Internet became more and more popular, the user base expanded from certain specific groups to the entire population of China. This also made Internet developers shift their focus from technology to user experience. With an accurate understanding of user needs and more improved services, a product can easily go viral. Unlike product development based on platform construction, "micro innovation" focuses on specific user groups and delivers the best user experience with meticulous attention to detail and continuous trial and error, in an effort to meet their specific needs.

"Micro innovation" is often considered to be the best methodology for Internet entrepreneurship in the age of mass innovation. A new product doesn't have to come from a new idea. With a single-minded focus on target user needs, and continuous improvement of the user experience, an entrepreneur can create an excellent product from scratch or based on existing products. WeChat is a typical "micro innovation" product. Before the WeChat project was launched, instant messaging apps for mobile had already swept the market. WeChat was inspired by Kik, an instant messenger app that was once top of the App Store. There were many imitators such as iGexin and MiTalk at the time. So how was WeChat able to stand out from so many competitors? First, its long standing presence in social media meant Tencent had an existing customer base that could easily be converted into Wechat users. Second and most importantly, the WeChat team's accurate understanding of user needs offered more streamlined and user-friendly messaging and social experiences.

WeChat has the following major "micro innovations":

Voice messaging. Although voice messaging is now widespread in mobile instant messengers, WeChat has been developing this function since the early versions, which clearly provides a better user experience than text messages, bringing WeChat a large number of new users.

Communicating with strangers. Location-based services (LBS) are one of the greatest advantages of mobile Internet. By integrating messaging and LBS for the first time in China, WeChat developed a feature called "People Nearby", so that users could communicate with strangers. They then launched another "micro innovation", a feature called "Shake", based on the gravity-sensing function of smartphones, ensuring easier and more interesting communication with strangers. With this new feature, WeChat experienced explosive user growth.

Clear social boundaries Instagram, a social media app based on photo sharing, has been amazingly successful. WeChat also integrated photo sharing into its 4.0 version and launched a new feature called "Moments". This thoughtful feature helps users create a clear boundary between strangers and their social circles. With "Moments" users and their friends can share photos, videos and text information if they belong to the same social circle. You can also contact strangers with the "People Nearby" and "Shake" features. This means it's great for communicating with friends and with strangers.

Protecting user privacy. WeChat protects user privacy in several ways, for example, "Moments" comments are only visible to mutual friends; user status such as "Online" or "Invisible" has been removed; and voice messages can be played in two different ways, depending on the distance of the phone from your ears. Users are bound to appreciate such careful design.

As the WeChat app was being developed, instead of creating original technologies or ideas, the company focused on improving user experience as much as possible with "micro innovations", continuously drilling down into user needs data and learning from existing technologies and products. They followed the principle of "Simple is Beautiful" and only offered the most user-friendly and practical features. They paid attention to every detail and applied iterative updates to enhance user experience. That's how WeChat managed to take the market by storm in such a short time.

With "micro innovation", it has grown into an unrivalled mobile instant messaging app; yet there is another reason it has become China's largest mobile Internet platform: the shift in Tencent's business development philosophy. Their previous development strategy took a rather aggressive approach, constraining the development of emerging companies by copying their products with extensive technological resources, and a large pool of loyal users. However, with WeChat, Tencent switched to an open and inclusive approach based on cooperation and mutual benefit, to build it into a platform for consuming information. A large number of creative features were launched after the WeChat Official Account Admin Platform was opened to the public. By investing in and opening up a second-level API to dianping.com, Tencent has made great progress in the area of lifestyle services. They also developed their e-commerce market after investing in and opening up a first-level API to JD.com. With the help of WeChat Pay, Tencent has expanded into the final phase of O2O business, making the WeChat ecosystem complete.

It has now become a strategic platform product of Tencent, with a strong competitive edge over Alibaba and Baidu. Their "micro innovation" approach to early development and their open and inclusive attitude after achieving success are valuable lessons for entrepreneurs.

References

1. Zhang X. Product concept behind WeChat. Forbes China, 25 July 2012
2. CNNIC Report Analysis. The emergence of Mobile Internet, Weibo, and Group Purchase. Sohu IT, 19 Jan 2011
3. Ma H. WeChat explores the international path for tencent. Tech QQ, 7 May 2013
4. Nan Q. What are the differences among WeChat, WhatsApp and Line. Tech Sina, 29 Jan 2015
5. Ma H. On WeChat platform: Set rules and engage third parties. Tech Sina, 7 May 2013
6. Zhang J. Communication changed by the internet after a decade, 80% interviewees communicate through WeChat. yicai.com, 26 Jan 2015
7. Zhang L. WeChat, the biggest winner of brand promotion on social media in 2014. www.qdaily.com
8. Sun L. Iteration and innovation of WeChat. 3 July 2014
9. Flash Sale of Xiaomi on WeChat. 150,000 Xiaomi3 phones sold out in ten minutes. Tech 163, 28 Nov 2013
10. War of Payment. Alipay V.S WeChat. 21st Century Business Herald, 23 Nov 2013
11. Nan Q. Three giants of social networking: what are the differences among WeChat, WhatsApp and Line? Tech 163, 29 Jan 2015
12. Zhang X. Eight rules for WeChat official account platform. Tech QQ, 11 Dec 2014
13. Toffler A (1980) The third wave. William Morrow, New York
14. Toffler A, Toffler H (2006) Revolutionary wealth: how it will be created and how it will change our lives. Alfred A. Knopf, New York
15. Ritzer G. Focusing on the prosumer: correcting an error in the history of social theory. In: Blättel-Mink B, Hellmann K-U (eds) Prosumer revisited. Verlag für Sozialwissenschaften, Wiesbaden, pp 61–79
16. Cochoy F (2015) Consumer at work, or curiosity at play? Revisiting the presumption/value co-creation debate with smartphones and two-dimensional bar codes. Mark Theory 15 (2):133–153
17. Ritzer G, Jurgenson N (2010) Production, Consumption, Prosumption: the nature of capitalism in the age of the digital 'prosumer'. J Consum Cult 10:13–36
18. Chia A (2012) Welcome to Me-Mart. Am Behav Sci 56(4):421–438

Case III: SHANGHAI GM: The Way to Intelligent Manufacturing

In the factory of Shanghai General Motors Co., Ltd. (hereinafter referred to as SGM), Jinqiao District, Shanghai, the leading role in the car body welding production line was rows of yellow robots. They kept busy from day to night, following instructions in order strictly and carefully, their arms stretching out and drawing back freely. After welding, original car bodys were transferred to the end of the production line one after another, where two operators took the responsibility for checking the achievement made by the robots. Together With those robots, the operators supplied one car to the next technique production line every minute.

Was this production line in SGM intelligentialized? "There's still a gap," one senior manager of SGM said, "The so-called intelligentialize means that the robot can make its own judgment and manufacture, but in our system, the robot's judgment is relatively limited. Now five or six car models are produced in the shared line. We set up different programs for robots to let them judge for different carmaking."

Intelligent manufacturing was the direction that SGM strove to work for. As on one hand, it should improve the production efficiency continuously, and on the other hand, it strove to satisfy the customer's more and more personalized customization demand. However, how would it change the present production model into the intelligentialize model, which was still a task for SGM. It needed to think about how to control cost in the model transformation, how to achieve intelligentialize through interaction between machine and machine and between machine and man, and how to improve the manufacturing efficiency through the virtual simulation.

The case is co-written by Prof. Xiaoming Zhu, Qiong Zhu, casewriter and Yifan Ren, research assistant in China Europe International Business School. During the writing, they obtained cooperation and support from SHANGHAI GM. The case aims at providing the source material for the class seminar but not illustrating whether the management of the company mentioned in the case is effective or not.

© Springer Nature Singapore Pte Ltd and Shanghai Jiao Tong University Press 2018
X. Zhu, *China's Technology Innovators*, Management for Professionals,
DOI 10.1007/978-981-10-5388-7_3

SHANGHAI GM

SGM was co-founded by SAIC Motor Corporation Limited (SAIC Motor) and General Motors in 1997. By the end of 2014, it had built four manufacturing plants of Jinqiao in Pudong, Dongyue in Yantai, Beisheng in Shenyang and Wuhan. There were 29 series products under three big brands of Buick, Chevrolet and Cadillac, covering various gradient markets from high-end limousines to economical cars, such as segment markets of high-performance limousines, MPV, SUV, mixed-powered and electric vehicles. In 2014, SGM achieved the sale of 1,760,158 passenger car vehicles for the whole year, ranking the second in the sale among the passenger car vehicle manufacturers in China. In 2013 and 2012, it achieved the respective sale of 1,575,167 and 1,392,658 passenger car vehicles, ranking the first in China.

From 1997 when the first pile for the factory in Jinqiao was driven to the time when the first car was finished in the assembly line, it took SGM 23 months. Being different from those self-owned brand car enterprises, from its birth, SGM could learn from the mature manufacturing and management models of American GM, its foreign partner. From then on, it began its information management system building and production automation construction firmly and steadily.

According to Leo Liang, CEO and Managing Director of Greater China of PLM Software in Siemens, the future road for the manufacturing industry in China (intelligent manufacturing) could start from building up the digitalization enterprise or factor.[1] Therefore, from its establishment, SGM set foot on the road to intelligent manufacturing.

Intelligent Manufacturing

What is Intelligent manufacturing?

Intelligent manufacturing was a man-machine integration intelligent system, which was composed of intelligent robots and human experts. In the manufacturing process, such intelligent activities as analysis, reasoning, judgment and decision-making were carried out. It expanded the manufacturing automation concept to be flexible, intelligent and highly integrated.[2]

In Hannover Industrial Fair in Germany of 2014, an "Intelligent Factory" co-developed by several German companies demonstrated the intelligent manufacturing scene to the public for the first time: on the exhibition booth, name card cases were manufactured in a production line. All information concerning manufacturing the name card case was input into the product components themselves through the

---

[1]Geng [1].

[2]"Intelligent Manufacturing", *China CNC Machine Tools*, 2006.05.17, http://www.c-cnc.com/news/newsfile/2007/7/15/260.shtml accessed 2015.03.02.

Table 1 Five features of Intelligent Manufacturing

First	Self-discipline. The processing machines or robots have the capability of collecting and understanding information, conducting analysis and judgment and planning their own behaviors
Second	Man-machine integration. Co-working, mutual "understanding" and cooperation equally between man and machine are achieved
Third	Virtual real technology. On the basis of computer, the signal management, cartoon technology, intelligent reasoning, predicting, simulation and multi-media technology are integrated; with the help of various audio and video and sensing devices, the manufacturing process and products were displayed virtually. It is one of the key technologies to realize the man-machine integration at the high level
Fourth	Self organization and super flexibility. All composing units in the manufacturing system can compose one best manufacturing unit independently on the basis of the task
Fifth	Learning and maintenance capability. The intelligent manufacturing system can enrich the knowledge base in practice constantly and has the self-learning capability. Meanwhile in its running process, it can diagnose errors by itself and has the capability of self-troubleshooting and self-maintenance

Source Lierun Rong, Intelligent Manufacturing toward the 21st Century, Mechanical and Electrical Integration, Vol. 4, 2006

Internet, and therefore those components could conduct information exchange with the manufacturing machine, commanding the machine "to produce me like this". Such scene, from the point of view of some experts, was just "a piece of cake" in the future intelligent factory. In the future, all machines, raw materials, transport vehicles and robots would make exchanges with each other and make independent decisions. Raw materials would contact with the processing machine. "Which machine should I go to for processing?" Then the processed work piece would tell the processing machine in the next working procedure "what materials I need". After that, in accordance with the underground induction line, the transport vehicle would deliver materials to the feeding robot. All subsequent producing procedures including manufacturing and sale documents were installed in those work pieces. In case some errors happened in the work piece or the customer had new requirements, the intelligent engineer in the R&D department made warning signals immediately and the improved measures after calculation were to be sent to the work piece.[3]

Compared with the traditional manufacturing model, the intelligent manufacturing had five significant features (see Table 1).

The intelligent manufacturing system was first proposed by Japan in 1989. In 1994, Japan launched the advanced manufacturing international cooperation study program including corporate integration and global manufacturing, manufacturing knowledge system, distribution intelligent system, distribution intelligent system technology for rapid realization of products and others. America, EC, Canada, Australia and other countries participated in the program.

---

[3]Peng [2].

Development at the State-Level Strategy

By 2014, the intelligent manufacturing had developed into a focus of competition in main countries globally. Traditionally developed countries and emerging countries positioned the intelligent manufacturing from the perspective of the state industrial structure reconstruction and the state competitiveness promotion.

Germany: In 2010, Germany formulated the ten-year automation development program, in which the manufacturing automation level raising was conducted as the state policy and the country would promote vigorously the development of electronic and electrical technology, mechanical and electrical integration technology, manufacturing technology process, computer and IT technology, sensors, driving and execution system, communication technology, comprehensive technology and others. In 2012, Germany unveiled *Ten Future Projects* program and Industry 4.0 was one of the ten future projects. In April 2013, Industry 4.0 was launched officially as the state strategy in Hannover Industrial Fair by the German government.

The term Industry 4.0 was proposed after the previous three Industrial Revolutions. The Industry 1.0 started in England in 1780 with mechanical production replacing manual labor; the Industry 2.0 took place in 1900, creating the product mass production model and the Industry 3.0 started in 1979, achieving the production automation and informatization.[4]

According to the viewpoints from Germany academic circles and industrial circles, industry 4.0 was the fourth Industry Revolution focusing on the intelligent manufacturing, aiming at transforming the automation into the intelligentialize in the manufacturing industry by the combination of the information communication technology and the cyberspace virtual system (Cyber-Physical System[5]).[6]

Industry 4.0 included the model transformation from the centralized control to the decentralized and enhanced control, aiming at setting up a personalized and digitalized product and service model with high flexibility. In this model, the traditional industrial boundary would vanish and various new activity fields and cooperation forms would emerge. Changes would occur to the process of creating new values with the industrial chain division restructured. Industry 4.0 contained three major tasks of intelligent factory, intelligent production and intelligent logistics.[7]

In accordance with the view taken by five doctors in German Chinese Mechanical Engineering Society, industry 4.0 was the Germany national strategy

---

[4]Balinski [3].

[5]Cyber-Physical Systems make the physical devices have the functions of computation, communication, precise control, remote coordination and self management through their connection to the internet. Through the cyber-physical systems, a virtual internet world is fused with a physical world.

[6]*Spimes, Cyber Physical Systems and Industries 4.0*, 360doc, 2013.08.05, http://www.360doc.com/content/13/0805/11/9561082_304859172.shtml, accessed 2015.03.03.

[7]Siemens, *Industry 4.0: Secure the future, grasp opportunities*, PACE, Feb 2014.

and it was not the international standard.[8] In fact, many countries had made such strategies such as America, Japan and Britain.

America: In June 2011, America launched officially the "Advanced Manufacturing Partner Plan" including the industrial robot technology. The Plan covered: first, $300 million was invested into the innovation of industries of small batteries with high power, advanced compound materials, mental processing, biological manufacturing and alternative energies; second, the material genome program, in which American enterprises would double their speed of discovering, developing and applying advanced materials through more than $100 million investment in the study, training and infrastructure. Third, the investment would be made in the next generation of robot technology. The next generation of robots would undertake works of workers, medical staff, doctors and astronauts. $70 million would be invested in this program. Fourth, the manufacturing technology, in which energy was used with high efficiency, would be developed.[9]

In February 2012, America introduced the "Advanced Manufacturing Industry National Strategy Program", proposing that the government investment be enhanced and the intelligent manufacturing technology platform be built up. One month later, President Obama tabled a proposal that $1 billion would be invested in building up the new internet of American manufacturing industry. The framework and methods of American intelligent manufacturing industry, digitalization factory and 3D printing technology were included in the development priorities.

Britain: In 2008, the financial crisis made the British government realize that the finance-focused service had failed to help Britain to continue to maintain its international competitiveness. Therefore, in December 2011, focusing on the "Advanced Manufacturing Industry Chain Proposal" program, Britain invested £125 million in the industries of automobile, airplane, renewable energy and low-carbon technology in order to build up the manufacturing industry chain. In January 2012, Britain started the strategic study of 2050 British manufacturing industry development. In October 2013, the final report *The Future of Manufacturing: A New Era of Opportunity and Challenge for the UK* was completed. The report pointed out that the manufacturing industry was no longer "the product sales after manufacturing" traditionally but the "service + re-manufacturing (the value chain with the manufacturing as its focus)". At the same month, Britain released its *British Industry 2050 Strategy*.[10]

Japan: As one of the early researchers of the intelligent manufacturing, Japan had been building up the intelligent system covering all industry chain. In 2011, Japan

---

[8]"Five Minutes Help You to Understand What Germany Industry 4.0 Is", *Gasgoo*, 2015.02.28, http://mt.sohu.com/20150228/n409218930.shtml, accessed 2015.03.02.

[9]"Partnership in Advanced Manufacturing' Program", W*enweibao*, 2012.04.17, http://news.hexun.com/2012-04-17/140472836.html, accessed 2015.03.02.

[10]"Britain Explores to Revive Manufacturing", *MIIT International*, 2014.12.23, http://www.up-tech.com.cn/news_infor.php?nid=711, accessed 2015.03.02.

released the Fourth Science and Technology Development Basic Program (2011–2015). In the Program, the R&D orientation was identified in multi-functional electronic equipment, information communication technology, measuring technology, precision machining technology and embedded system from the aspect of industry with sophisticated manufacturing layout and to improve the intelligent manufacturing support technology of the intelligent network, high-speed data transmission and cloud computing at the same time. From 2013, Japan had enhanced such frontier technology research as 3D printers. In 2014, Japan invested ¥4.5 billion (about ¥235 million) to implement the R&D program of "Product Manufacturing Revolution with 3D Modeling Technology as Focus" and hoped to develop the most advanced 3D printer for the metal powder modeling.[11]

China: Before 2014, Chinese government and enterprises had been working on the integration of industrialization and informatization. In August 2014, the Minister of Industry and Information Technology in China mentioned, "it is the urgent priority to promote the entire industry transformation and upgrading with the internet by means of the intelligent manufacturing." He thought that the era of the equipment manufacturing industry with an annual high-speed growth of 25% came to an end and it was urgent to take the high-end equipment manufacturing as the intelligent manufacturing as a leading role to promote the entire industry transformation and upgrading. So the Ministry of Industry and Information Technology would push to develop such key components and devices as the industrial robots, sensors and intelligent instruments and particularly advance relevant standard system construction.[12] In October that year, the Premier Keqiang Li signed industry 4.0 cooperation agreement with German.

Intelligent Manufacturing Development in Automobile Industry

One of the major roles of the intelligent manufacturing was the intelligent robot evolving constantly from robots and the automobile industry was the first to use robots. In 1961, the first industrial robot showed up in the production line of American GM, the "master" of SGM.[13] Then the industrial robots all over the world had been co-developing with the automobile industry. For example, Japan's YASKAWA, FANUC and Toyota and Honda, FANUC and GM, German KUKA and Volkswagen, Italian Staubli and Fiat were the long-term cooperation partners.

---

[11]Zhang [4].

[12]"China Will Promote Standard System Construction of Intelligent Manufacturing and Internet of Things", *Zhongziwang*, 2014.08.28, http://www.wotchina.cn/news/zhengce/2116.html, accessed 2015.03.02.

[13]Xinci and Xuena [5].

The analysts predicted that from 2012 to 2016, the average annual growth rate would be 4.81% in the global automobile robot market.[14]

With the robot application, man-machine collaboration was the scene in some automobile factories. For example, in the Volkswagen engine manufacturing plant in Salzgitter, Germany, in a work procedure, the glow plug should be inserted in the almost invisible drilling hole of the cylinder cover. Previously, in this work procedure the worker bended hard to eye at the hole whose job was now transferred to the robot. The robot inserted the glow plug into the drilling hole and the worker just conducted the insulation treatment for the cylinder cover by fixing the glow plug. With the help of the robot, the workers did not need to bend repeatedly and they could finish the work procedure by standing there relatively comfortably. Meanwhile, they could also monitor the whole process and carry out intervention if necessary to ensure the smooth production. "We make use of a workplace layout that is consistent with ergonomics so that all employees are freed from the long-term burden in the company. By using robots without the safety fences, employees could work with robots together," said the project manager in the plant. Such man-machine collaboration helped the plant optimize the production line so that the average consumed energy and pollution emission reduced by 67 and 70% respectively for the one engine production.[15]

The robot application resulted in the production automation so that some automobile manufacturers began to try out the collaboration management even the intelligent management of the automation production. In a new plant manufacturing the BMW3 Series in the suburb of Leipzig, Germany, the collaboration management concept was introduced in the plant workshop design. There was a central building with the area of 26,000 m^2 in the plant, and the car body workshop, painting workshop and general assembly workshop were built up around it. The star-shaped building complex layout kept the manufacturing distance short. On the ceiling of the central building a free hanging conveyor system was installed. Through the system the original car body was delivered to the car body storehouse from the car body workshop, then to the painting workshop and the painted car body was delivered to the car body storehouse again. After that, it was delivered to the general assembly workshop from the car body workshop. The central building played not only a role of the coordinator of the logistic transportation but also the role of the collaboration management for the automation production line composing of robots and men. Since the parts of each BMW3 Series car were placed internally in the RFID chip[16] for the individual information identification, therefore with the

---

[14]*Research and Markets Adds Report: automobilemotive Industrial Robotics Worldwide Market Report-with Forecast to 2016*, Professional Services Close-Up, Nov 8, 2013.

[15]Rapid Science and Technology, "German, Intelligent Robots Make Interpretation of automobile Industry 4.0", *Cheyun*, 2014.12.31, http://news.mydrivers.com/1/363/363180.htm. accessed 2015. 03.02.

[16]Radio-frequency identification (RFID) is a technology to identify and record the presence of an object using radio signals.

help of the information, the system could match the vehicle with the corresponding processing equipment, technology and material.[17]

The application of robots was not new for the automobile manufacturers in Chinese market. In all new entire car production lines, regardless of self-owned brand or joint-ventured one, the fully automatic production line with robots were introduced.[18] Some enterprises were transforming old production lines with robots gradually. Moreover, the automobile enterprises like Chery conducted research and development of robots independently. After satisfying its own demand for the automatic production, robots in Chery were also sold to JAC Motors, Midea Aircon and other enterprises.

However, in order to make those robots intelligent and help robots and human experts make intelligent analysis, judgment and decision in the manufacturing process, the digitalization or informatization foundation should be laid during the entire manufacturing process, including the digitalization operation, networking and virtual reality. From the first day of its establishment, SGM carried out practices in those aspects.

SGM: Informatization

On June 12, 1997, the information system department was established when SGM was founded, planning and setting up independently and comprehensively the computer office automation environment, SAP-based accounting, procurement management system, flexible manufacturing application system, automatic tracking application system in manufacturing vehicles, quality management application system, equipment technology monitoring application system. Furthermore, such global application platforms as American GM's product data application platform, sale application platform, material demand management platform, dynamic cost accounting management platform were built up.

At the beginning of its establishment, in the light of the fact that most employees were graduates just out of schools and they lacked knowledge about the automobile manufacturing and management, SGM decided to deploy American GM's global application platform for some key business managements. Therefore, on one hand, the information system could be built up rapidly to satisfy the overall construction progress of the company and start production as scheduled. On the other hand, through learning the system to get to know the business, the company's business would be developed.

On April 12, 1999, the first Buick car was completed on the general assembly line and then SGM entered the track of rapid development. GM's global application

---

[17]"BMW Leipzig Plant Adopting Collaboration automobilemation Concept", *Siemens (China) Co., Ltd.,* 2010.11.04 http://www.zgznh.com/fangan/show-713373.html. accessed 2015.03.02.
[18]Xinci and Xuena [5].

system was operating in the large mainframe which were placed either in the northern America or in Europe. Due to the time difference, while SGM was working, those mainframes were under the system maintenance. Moreover, the network lines were not under control, and consequently the system became unstable so that SGM was very passive. As a result, the more or less trouble was hidden to bring about adverse effect to the Company's normal business development.

The year 2000 was an inflection point that demand changed in Chinese automobile market. The main automobile demand with the car and mini bus changed from the group consumption into the private consumption. Two years later, the "blowout" showed up in the Chinese passenger car vehicles market. In 2002, the market grew up by 36.65% year on year.[19]

In such a market, in order to provide better services for customers to develop market rapidly, SGM built up the customer relationship management application system in 2001.

However, in the global sales system, the rushing for the higher sales volume always happened at the end of the month, and the application system did not work properly time and time or the network line was out of service, and as a result the sales business was effected adversely many times. Therefore, the information system department decided to build up the localized sales system to make the sales business operate nomally. In 2002, the system was on line for service.

One year later, SGM's dealer management system offered its service on line. In Chinese automobile industry, the system created a precedent for the dealers' information management.

Having experienced the effects brought about by those local system, SGM came to understand profoundly the importance of the key application system localization. Therefore, on the basis of the Company's specific demand, it carried out a comprehensive study of the alternative and applicable commercialized software. After one-year preparation and half year comparison and test authentication, SGM started to build up the comprehensive localization application platform in 2003.

From 2005, on the basis of its business demand, SGM set up vehicle maintenance data analysis platform, the entire car sale predication system, the spare parts prediction system, the spare parts inventory management system, the customer dimension analysis, the assets management system, the accounting forecasting system, the media application management system and the vehicle electronic development platform, etc. By the mid year in 2006, except for the product data application platform, all SGM application systems were set up on the basis of SGM'S automobile data center. Then the data disaster recovery center was established in Yantai, Shandong.

There are three questions for manufacturing enterprises: what to produce, how much to produce and how to produce. The first two questions were involved in SGM's enterprise resources planning system, supplier management system, but

[19]"Why 'Blowout' Showed up in China's automobile Market in 2002" *Economic Information Daily*, 2003.03.11.

how was the relevant information on "planning" broken down and assigned to the "production execution" line? How was the status fed back rapidly to the "planning" link in the "production execution" process? Most Chinese enterprises had solved this problem by passing the plan to the workshop manually. Thus the ability to response to the market and improve the product quality was effected adversely.

SGM built up a bridge between "planning" and "producing"—MES solution. MES was a set of production information management system for the execution level in the workshop. Through the information transmission, MES could conduct the flexible manufacturing management from placing the order to completing the product in the whole production process.[20] In SGM, the MES solution covered the on-site order management system, material release system, production monitoring system and production control system. Through those systems, the mutual and timely interaction was achieved in the business management and production.

SGM: Automation

Like informatization, initially the automation in SGM copied from the one in American GM. So in the former car body workshop in Jinqiao, the welding robots were used. Besides, robots in the painting and general assembly line were introduced in succession. Those robots had experienced upgrading for four times. In 2015, the fifth generation of robot was introduced. According to the relevant person in charge in SGM, compared with the previous generation, the new generation of robots ran faster with smaller size, cheaper price and higher efficiency.

Reducing the labor cost and improving the automobile quality were the first motivation for SGM to use robots. SGM's robots supplier was FANUC—one of four top suppliers in the global industrial robots field. American GM signed a strategic cooperation agreement with FANUC, so SGM could buy robots at the negotiable prices. Robots not only reduced the cost but also improved the product quality. For example, the assembling and welding technology assured the position standardization of thousands of welding points and their strength in the entire car.

Having experienced various advantages of robots in Jinqiao plant, SGM took the robot production line as the standard technology in other plants that were purchased or built newly by it. The robot model, system structure and manufacturing technology adopted by each factory complied with GM's global standard. For example, in the car body workshop built up at the third phase of expansion project in SGM Dongyue plant, there were 236 robots, which could complete processing one car within 51 s. Furthermore, robots increased to 369 in the car body workshop at the third phase of expansion project in SGM Beisheng plant, and the production line could accommodate the co-production of five different models with the automation

---

[20]"Information Technology Fault in MES Manufacturing", *Hill Technology*, 2008.05.25, http://www.systron.com.cn/mes20.htm. accessed 2015.03.02.

rate of 93%. The index referred to the percentage of the welding point numbers by robots out of the total welding points in the car body. In Jinqiao plant, the automation rate reached 85%. In addition, the automation rate arrived at 97% in SGM Wuhan plant, which was completed in January 2015 and where there were 465 robots in the car body workshop.

To manage those robots who were working actively in the production line, by the end of 2014, SGM still had adopted the manual point-to-point model, which meant that one man carried out his management of one robot by entering information through the interface. Nevertheless, it had noticed the development orientation toward the intelligence technology. The network function was installed in the third generation of FANUC robots. 60% of the third, fourth and fifth generation of robots owned by SGM was installed with the network function as well. Since the software centralized management and monitoring functions by the third party were not perfect, SGM had not realized the centralized management of robots through networking, but the relevant person in charge in SGM said that the tendency of the intelligent networking management was obvious and it had been doing some relevant preparations. For example, SGM was exploring the possibility of replacing practical work with virtual simulation. Its robot offline simulation study started in 2010. SGM hoped to complete verification of the robot machines, the project commissioning and the processing technology. By the end of 2014, its coincidence degree of the robot offline simulation had reached 70–80%, which meant the on-site workload was reduced significantly. The person in charge also said that as soon as the relevant software was improved, the offline simulation procedures would play a greater role in the network environment.

In the automobile industry, the manufacturing changing from automation to intelligentialize was a new task for almost all enterprises and SGM was no exception. On the basis of its own informatization and automation, what a track would SGM leave on the way to the intelligent manufacturing?

Case Analysis I

Intelligent Manufacturing Ushers in the Era of Industrial Big Data

Dong Ming

Intelligent factories represent a modern factory model where digital technology is used to integrate knowledge, intelligence and staff experience in the fields of product design, manufacturing processes, production management, corporate management, sales and supply chains. This in turn supports product design, production,

Dong Ming, Operations Management Professor and Associate Dean of Antai College of Economics & Management at Shanghai Jiao Tong University.

management, sales and services. The model depends heavily on Ubiquitous Networking (Internet and IoT) technologies to obtain real-time data and information both inside and outside the factory, optimise all the activities of production units, maximise efficiency in production, logistics and resource use, protect the environment and unleash the potential of labour forces.

Intelligent factories, intelligent production and intelligent logistics are the three major concerns of Industry 4.0. Intelligent factories focus on intelligent production systems and processes, as well as networked and distributed production facilities. Intelligent production mainly involves corporate-level production management, human-machine interaction and the application of 3D printing technology in industrial production. Intelligent logistics integrates resources via the Internet and IoT to improve the efficiency of existing logistics resource suppliers and provide real-time logistics support for customers. Intelligent manufacturing is prioritised by large companies such as SHANGHAI GM (SGM) in their future development plans and attracts small and medium-sized enterprises so that they can benefit from next-generation intelligent production technology and contribute to the innovation and supply of advanced industrial production technologies.

Intelligent factories have three main aspects: engineering, production and supply chains. This description is based on a 3-part information model: engineering (design-centric), production and manufacturing (management-centric) and supply chain (raw material and product sales). This case study explicitly introduces the information-based transformation and construction process of SGM, making information transformation and automation a strong foundation for SGM on the path to intelligent manufacturing. Indeed, the IoT and service networks are the information and technology foundation for intelligent factories. The service network is closely linked to enterprise resource planning (ERP) and customer relationship management (CRM) for production planning, logistics and operation and product lifecycle management (PLM) which is used for product design on the top layer. The industrial IoT is closely linked to the cyber-physical system (CPS) which enables manufacturing execution systems (MES) to produce production equipment, control production lines, manage production and perform advanced planning and scheduling.

Inspired by the German Industry 4.0 strategy, China recently proposed the "Made in China 2025" strategy, a three-stage approach for transforming China into a world manufacturing power in 30 years, which will serve as the action guidelines for the first ten years. To ride the wave of the new technological and industrial revolution, China proposed the "Made in China 2025" strategy based on actual demand in the changing economic growth model. The strategy focuses on key processes such as innovation-driven development, intelligent transformation, strong foundations, green development and staff development. It also prioritises key sectors, such as advanced manufacturing and equipment, with the aim of accelerating manufacturing restructuring and upgrades, completing major strategic tasks and implementing policies and initiatives that can increase efficiency and transform China into a manufacturing power which emphasises quality over quantity by 2025.

The case study clearly explains the development path for SGM towards intelligent manufacturing and finally, the article raises a question about SGM's future: On the basis of its own information transformation and automation, what path will SGM take on the way to intelligent manufacturing? One possible answer could be: industrial Big Data. In the intelligent manufacturing era, industrial machinery, equipment, storage systems and operational resources can be connected through network communication technology. Information can be shared between factories, machines and equipment at any place or time and the interconnected system can be independently self-managed. So the transformation to intelligent manufacturing will usher in the era of industrial Big Data. This data is generated by intelligent industrial systems or product manufacturing processes. As the output of intelligent manufacturing, this data is an indispensable condition of intelligent manufacturing.

Case Analysis II

SGM: Creator or Follower?

Li Yuan

Looking through the SGM case study, I see there has been technological advancement, and improvement in quality and efficiency, but no innovation.

I think innovation means creating a new product and breaking into new markets, like when the emergence of smart phones disrupted Nokia overnight; or adding value, for example, the Beijing Youth Daily paper boy would take the subscribers' rubbish away; or delivering a new user experience, like when Delphi pioneered radio installation on GM vehicles; or increasing efficiency, for example, through the use of robots, specifically the people who break new ground in using them, not those who follow afterwards.

Although the case study seems to show that SGM is taking the lead in automation, intelligence and Big Data, in fact it is following the trend. In the car market, all car manufacturing companies are transitioning towards intelligent manufacturing. If they don't make this step, their poor quality and low efficiency mean they will be replaced by their competitors. The case study clearly shows that SGM use the same practices as their parent company GM in North America, with certain adaptations to meet local needs, for example, SGM use a local server as the response time from North America was too slow. But generally speaking, SGM just follows GM's lead. If the headquarters accelerate their innovation, SGM follows and may take a lead in China, but then if innovation at the headquarters slows down, SGM also slows down and may even lose its edge in a developed market like

Li Yuan, student of EMBA 2011 of CEIBS, Director of the Asia-Pacific Safety Electronics business in Delphi.

China. So I think that SGM simply takes the necessary steps to maintain basic competitiveness, but this cannot be defined as innovation.

Based on this case study, I'd like to discuss a further possible innovation scenario.

As demand surged in the Chinese car market after 2000, SGM developed a dealer management system to effectively predict market demand and determine the number of cars to be produced and sold within a certain period. No company in China had done this before. However, when all the other car companies followed suit, both international brands and local brands once seen as cottage industries, SGM lost its edge. Another aspect is service. When customers place orders in 4S stores, they state their preference for the colour of the car body, the choice of fabrics, the colour of accessories, etc., and then they wait one month to get their cars. Yet they often get impatient or frustrated by this process, and they would be willing pay to get their car earlier, compromise on customization or let it go. The customer experience is terrible. If more personalized configurations could be added that do not require precision equipment on the final assembly line/, or if suppliers offered accessories that could be installed manually, these accessories could be sent to the 4S stores directly and car manufacturing companies would only need to produce cars with standard configurations. This would reduce production time per car and the workload of special equipment, and manufacturers could stock fewer accessories, while increasing the speed of car production and offering a faster response time to customers. Customers could probably get their ideal car within days or even hours of visiting a 4S store. To achieve this, car manufacturing companies would need to change their design, improve supplier management, and transform the sales process, business model and financial model.

Ensuring the most simple approach to production and operation is also a form of innovation. Take lean production, a concept proposed by Toyota 20 years ago. Japan pioneered automated production focusing on quality and precision. Its product quality was the best in the world. At that time, technological limitations meant that spare parts of certain shape or size could only be produced through an expensive production process. Japanese companies solved the problem effectively by introducing manual operations. For example, Toyota's motorbike frames were sometimes a little distorted after the shaping and welding process and adjustment with automatic equipment would be very expensive. So they introduced a metal detection tool to lay against the surface of the finished products and if the surfaces did not fit perfectly, they would be hammered into shape. If you visit the factory floor of any leading car brand, you'll find that as well as the advanced robots, there are many "skilled staff" working on the final assembly line. They punch, kick and push the car doors and trunk lids that are not well assembled and these manual adjustments are not performed with any detection tools. Intelligent automation does not necessarily lead to innovation, but rational simplicity for lean production can produce innovative outcomes.

Objectively speaking, SGM definitely excels in certain aspects. Although I have described SGM as a passive follower, in fact, they are making great strides in terms of localization. For example, with the growing trend for mobile Internet and the

Internet of Vehicles, many international car brands look to Google, while SGM clearly understands the advantages and market position of Baidu in China, and has taken the lead in adopting Baidu Car Life in the information and entertainment systems of its advanced cars. Though the future is hard to predict as followers are catching up, the combination of vehicles and the Internet offers a fresh new customer experience, with a blend of international practice and localized versions.

References

1. Geng J (2013) "Siemens PLM Leo Liang: The Future Manufacturing Starts from the Digitalization Factory", ChinaByte, 2013.12.19, http://soft.chinabyte.com/390/12810390.shtml . Accessed 02 Mar 2015
2. Peng X (2015) The Industry 4.0: from automobilematic Manufacturing to Intelligent Production. Big Aeroplane (2)
3. Balinski B (2013) *German companies developing Industry 4.0*, Manufacturers' Monthly, Apr, 2013
4. Zhang Z (2014) "Artificial Intelligence Leads Japan to Seize Height of Advanced Manufacture", China Security, 2014.12.19, http://www.cs.com.cn/hw/hqzx/201412/t20141219_4594728.htm. Accessed 02 Mar 2015
5. Xinci W, Xuena L (2015) "Chinese-styled Man-Machine Collaboration". Financial News New Century (8)

Case IV: Can Robots Raise Laying Hens?

In the early spring, the West Fangezhuang Village in the northeast of Beijing was very quiet. When the car turned into a path, one row of gray buildings after another were standing in front of us in a wide open-land suddenly. If there was no reminder of the plaque written with "Beijing CP Eggs Co., Ltd." at the gate, it was hard for visitors to imagine that this was a hen farm. Here no cluck of hens and roaring of machines could be heard, and it was as quiet as places around it. Except for two security staffs at the gate, no one else was seen. A few of trucks parked beside a row of buildings.

In the building with the area of 779 mu (1 mu = 0.0667 hectares), the residents with the largest population were hens, annually 3 million of which were on hand, and the least population were people. Here 18 enclosed henhouses were seen, each with 127 m in length, 17 m in width and 9.5 m in height with constant temperature, constant humidity and good air quality. Every henhouse was equipped with one feeding engineer. The engineer and two robots undertook the responsibility for managing nearly 170 thousand hens. Each henhouse consisted of two floors, and each floor was inspected by one robot. The robot was patrolling through seven passageways in six rows of hencoops. The detection devices were installed in the robot's head, chest and knees. As it moved along rows of hencoops, the data of each chicken's temperature and the temperature and humanity around it were captured and transmitted to the central control computer (see Fig. 1: Robots Were Working in Henhouses). Through the computer, the feeding engineer carried out the real-time control of all hens' conditions in the henhouses and took counter measures if necessary. For example, he took the hen-to-be-eliminated out of the hencoop or submitted a report for repair out of the machine failure or just maintained it simply.

The case is co-written by Prof. Xiaoming Zhu, Qiong Zhu, case writer and Yifan Ren, research assistant in China Europe International Business School. During the writing, they obtained cooperation and support from CP Group. The case aims at providing the source material for the class seminar but not illustrating whether the management of the company mentioned in the case is effective or not.

© Springer Nature Singapore Pte Ltd and Shanghai Jiao Tong University Press 2018
X. Zhu, *China's Technology Innovators*, Management for Professionals,
DOI 10.1007/978-981-10-5388-7_4

Fig. 1 Robots are working in henhouse. *Source of Picture* CP Eggs

The feeding engineering worked here for eight hours per day, but the robot worked all day except for being charged.

Although men were the minority in henhouses, Shancheng Wang, vice president of Beijing CP Eggs Co., Ltd. (hereinafter referred to as: CP Eggs) tried to cut down employees, and he hoped that one man could manage three henhouses and more jobs were to be given to robots. Why did he lay off employees? Could robots undertake the job of raising hens? To realize the raising hens by robots, what preparation should this enterprise make?

Beijing CP Eggs Co., Ltd

Beijing CP Eggs Co., Ltd. was founded and wholly-owned by Thailand CP Group in 2010 in order to run and manage the hen farm in Beijing. CP Group was founded by the Thai Chinese in 1921 and built up a fortune out of the crop seeds sales. CP Group developed gradually into the whole industry chain management of agriculture and animal husbandry such as planting, feed, farming, agriculture and animal husbandry products processing and food sales.

Beijing Farm was a pilot project of CP Group in China to realize the automation, intelligentialize and industrialization in hen farming and egg producing, including the hen raising farm, chick raising farm,[1] feed processing farm, egg products grading and packing and liquid egg processing. Among them, there were three million hens in the hen raising size, producing 2.3 million fresh eggs per day and 54 thousand tons

---

[1]Young chickens refer to minor chickens at the age of 7–20 weeks that are nurtured to be mature sexually after leaving the warm house.

of fresh eggs per year, which accounted for 18% of the egg demand in the Beijing market. There were one million chicks with the annual output value of 1 billion yuan. The automatic farming and production line in the entire farm were all introduced from the professional machine production company in developed countries.

It was the idea to "rebuild China's food safety Great Wall" that made CP Group to build up the farm. The later was one hope of Dhanin Chearavanont, the President of CP Group. Therefore, the farm not only adopted the closed production model during the whole process from the feed processing to egg products transported out of the farm. It was also hoped that during the whole production process, the no-man contact should be done as much as possible. The construction of the farm commenced in April 2010 and was completed in April 2012. In July that year, the first group of chicks was brought into the farm. Four months later, they grew up into layers and moved into the layer farms. 18 henhouses were fully occupied in July 2014. During the process, robots got to work one after another.

Why Did CP Eggs Let Robots Raise Hens?

Safety

Replacing men with robots was decided when CP Group planned this project. The aim was to ensure the safety of hens and eggs. The food safety issue was the focus that the whole society showed its concern with, and furthermore, it also contained the tremendous business opportunities. Meanwhile, in CP Group, the quality safety issue was the red line. If the manager stepped on the line, he would be fired no matter who he was.

Wang had worked in CP Group for nearly 20 years. From his point of view, the greatest risk for raising hens was the biological safety risk and the risk's major source was man. Man came in or out the henhouse, so it could not be a completely-closed space. In addition, man might bring about bacteria into the henhouse. At the stage when the henhouse had to be managed by man, in order to prevent the risk arising from man as much as possible, the feeding engineer should wash the head, take a bath and put on work suits every day before he came into the henhouse. Moreover, the feeding engineer was allowed to go out of the farm twice per month. Every day when he went off work he should be back to the dormitory, his dormitory was isolated from the dormitories of the employees who did not contact with chickens. Obviously, such requirement for the feeding engineers was not persistent, so fewer people should stay in the henhouse and this was the direction for GP Eggs Co., Ltd. to strive for.

Cost and Efficiency

In the comparison with the ordinary employees, the management cost for the feeding engineers in CP Eggs was higher. Some employees joked that managing the feeding engineers was like managing pilots, but Wang said that this was a small sum of investment. From his point of view, the cost for those engineers' pay and training was not a small spending which furthermore was growing up more as the labor cost was increasing in China. "The pay for those engineers in our company was highest compared with employees in the posts at the same level, with the pay of at least 10,000 yuan per month." Wang also said that in addition to what just mentioned, the spending for training was also required for investing. Those engineers were mostly undergraduates or graduates, because the engineers in those posts should have strong learning ability. The engineer in the henhouse not only made use of the information technology and intelligent means to manage 170 thousand of hens but also distinguished between the healthy and the sick hens beside the hencoop and got the sick ones out of the hencoop. Such inter-disciplinary talents with the wider professional knowledge span were not ready-made, so CP Eggs needed to make investment in training such talents. However, those talents might not stay in CP Eggs after they were trained, with respect to which Wang said that 120,000 yuan would be saved in wages annually if the robot could be used to replace one feeding engineer.

Furthermore, Wang said that replacing men with robots could help improve the general production efficiency, because hens that robots raised could stay in a stable and comfortable environment, their egg laying rate and quality would be improved. Visitors of CP Eggs looked at hens through windows at the end of the henhouse in the corridor for visiting separated from the henhouse, but according to one employee's view, some hens were frightened to lay eggs with white shells when more visitors came, while normal eggs with maroon shells.

As of March 2015, the efficiency comparative advantage of using some robots to raise hens showed up in CP Eggs whose chicken breeds were introduced from German Lohmann Poultry Breeding Co., Ltd. The egg laying rate reached 103% in February, 3% higher than Lohmann hens in other places all over the world with large-scale farming. In addition, its henhouse environment ensured 99.8% of survival rate of layers, while the survival rate of layers in the other large-scale farms was around 80%.[2]

Customized Robots

Robots which were less than 30,000 yuan in CP Eggs were not bought and were co-developed by CP Eggs and a Chinese company. The reason that CP Eggs preferred the joint development model was that they failed to find one robot who could raise chickens when they travelled the world. Because this was a blank market, the Chinese company was willing to conduct co-development with CP Eggs.

---

[2]"Reflection on Development and Changes in the Layer Industry at the Transition Period", *Beijing Huadu Yukou Poultry Co., Ltd.*, 2013.02.16, http://www.hdyk.com.cn/newsimginfo.aspx?m=20120925090736950227&newId=20130216160722180461, accessed 2015.03.12.

Because of the requirement in food safety, cost and scale efficiency, constant penetration of the internet, and other various types of technology and new machines into the poultry farming, the large-scale, automatic and even intelligent chicken farming had become one development tendency in Chinese market.[3] Such a tendency had become or was becoming reality in some countries all over the world. For example, in America, the quantity of layer farming was 286 million, and there were 64 enterprises with more than one million layers, which accounted for 85% of the total layer farming quantity throughout America, and there were 11 enterprises with more than 5 million of layers. In Japan the quantity of layer farming reached 181 million with 4550 households and each household's egg laying quantity was 39,800.[4] In 2013, more than 1 billion layers scattered in 770,000 egg farms with the average amount of layers on hand of less than 2000 in China and there were only 4–5 farms with the amount of layers on hand of above 500,000.[5] Therefore, there was a huge development space between the reality and the tendency in China.

In this development, robots had been playing a more and more important role. In accordance with data offered by International Federation of Robots (IFR), by the end of 2013, there had been 40,000 milking robots all over the world. In 2013, the global sales of milking robots were 5100, with the 6% growth year on year; the sales of robots in the poultry farming such as cleaning robots or automatic grazing robots were 760 sets, with the 46% growth year on year.[6] In addition, robots conducted fish farming in Japan, and emulating sows replaced sows to feed piglets in Canada, and in France there were nursemaid robots for chicks and chicks could be taken out in the hatching site to be vaccinated.

The concrete implementation of the program of chicken farming by robots was carried out step by step after the farming buildings were completed and even layers were raised. For example, CP Eggs should design the robot's patrolling routes in accordance with the actual henhouse layout. An employee said that "in order to teach robots how to patrol properly, employees spent half year." The employees' requirements for robot's function were refined, standardized and upgraded as the layer automatic production line came into operation and they came to be familiar with the management works gradually and work out some laws. As of March 2015, they entered the second stage of the robot application. At the first stage, robots could read only data of temperature and humidity wherever they patrolled; at the second stage, robots could measure the temperature of chickens actively by relying on the infrared equipment. If the chicken's temperature was higher or lower than the normal value by 41.5 °C, the robot could make a judgment that this was an

---

[3]"The Future Tendency out of the Present Poultry Industry Development in China", *Veterinary Drugs Market Guide*, 2012.05.04, http://www.boyar.cn/article/2012/05/04/429510.2.shtml, accessed 2015.03.11.

[4]Han [1].

[5]Zhong [2].

[6]"IFR Releases 2014 Robot Study Report: The Market Is Rising Significantly", *Robot Portal*, 2014.10.08, http://www.roboticschina.com/ART_8800694440_500001_NT_63833721.HTM, accessed 2015.03.12.

abnormal chicken and sent the positioning information of the abnormal one to the central control terminal in the henhouse.

"We are still in the constant exploration on the development of the robot's function. The possibility of the function development depends on the development of the robot technology on one hand and our ability to propose demand on the other hand," Wang said. "The ability to propose the demand depends on our familiarity with this set of the business procedures of modern layer farming."

Those robots were still at the stage of experiment and exploration in the farming production line in CP Eggs, but Wang believed that his own production line enjoyed certain intelligentialize compared with those automatic hen farming production lines. He took the temperature and humidity control in the henhouse as an example: in most henhouses, the temperature and humidity were controlled by man on the basis of the set temperature, so when the temperature was high, the ventilation started. But in CP Eggs, the man was not involved. The computer would made independent judgment and decided to take necessary measures on the basis of various parameter. Whether the ventilation should start or not depended on not only the temperature but also the contents of ammonia and carbon dioxide in air. Meanwhile, the temperature difference between the inside and outside house was also taken into consideration. So the henhouses remained in a comfortable environment adapted with the reasons. Most parameter data that were required for the computer's decision-making were supplied concerning all information points in real time by the robots. It took four hours for the robot to finish its patrolling in one henhouse, so any information about every point was obtained every four hours.

Those data not only supported the intelligent decision-making in the farming production line but also provided the basis for the decision-making about the industrial chain extension by CP Group. For example, they had set up a crocodile farming supporting company on the basis of layer elimination rate. The plant was only five or six kilometers away from CP Eggs with more than 1000 crocodiles bred there. The plant made profits by selling crocodile meat and skin. In addition, the manure from layers and chicks in CP Eggs were about five tons annually. Wang believed that the manure could be processed into the high quality organic fertilizer, because their chickens were very healthy and the chicken's feed was the top-quality products processed by CP itself. Thus, they were planning to build up a supporting organic fertilizer plant. By making use of the organic manure, they would construct the organic peach forest and vegetable garden that were enclosed around the chicken farm of CP eggs and invited local peasants to participate in construction and management.

Innovating Business Model

The hen farming project of three million layers, according to Wang's view, had bright future, however, the early investment of 720 million yuan erected a high threshold for the project immediately. Even though other enterprises wanted to imitate it, the threshold could not be crossed easily. In fact, without innovating the new business model, CP Group would have stopped before such a high threshold.

Four-in-One Model

The business model created by CP Group was the "Four-in-One BOT (Built-Operate-Transfer) Property Agriculture" model (hereinafter referred to as Four-in-One Model). In the model, four meant: government, bank, leading enterprises and professional farmers' cooperatives. Four-in-One meant that through a certain mechanism, four parties were integrated, and with a completely-new organization model, dealing with problems and challenges arising from the development. BOT Property Agriculture meant that farmers made scattering land circulate into a single block of land in the manner of establishing cooperatives and this block of land could be used for the whole project. The farmer cooperatives leased the land to the project and obtained the land rent and other return on assets. While the government and leading enterprises set up the financing platform jointly to settle the funding problem for the project and made investment in construction as well. The well-established assets were authorized to the professional management company under the leading enterprises to carry out the business management and take risks. The management company returned 10–12% of the total investment amount as profit to the financing platform annually and the platform repaid the loan and paid the land rent to farmers (see Fig. 2).

As the model was adopted to the hen farming project specifically, CP Group jointed with the government of Pinggu District in Beijing to found a joint venture titled "Guda Agriculture Investment and Financing Platform" (hereinafter referred to as "Guda"). The platform undertook the responsibility for financing and carrying out the project construction; "Guda" went to Beijing Bank and took it as the major

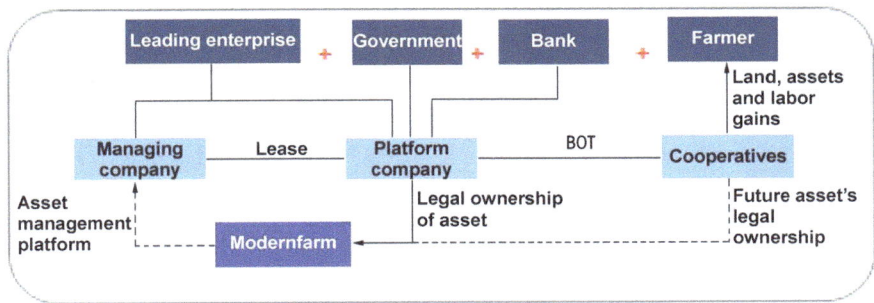

Fig. 2 Diagram of Four-in-One Model. *Source of Picture* CP Group

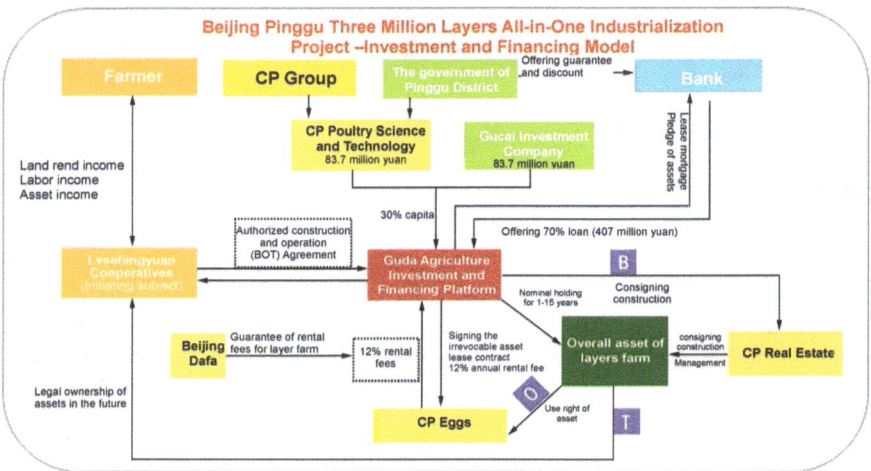

Fig. 3 Diagram of Pinggu Project. Adopting Four-in-One Model *Source of Picture* CP Group

financing source; the land for the project was the land owned by the cooperatives funded by 1416 farmers. As the project was completed, CP Eggs Co., Ltd. assumed the responsibility for the management (see Fig. 3).

Innovating Process of the Model

The model innovation was proposed, upgraded and improved gradually for a series of problems arising from the implementation of the three million layers farming project.

Solving the Land Problem

The land trouble occurred as the project started: how to obtain 779 mu of land conveniently? As Notification Concerning Promoting Relevant Land Policy on Large-Scale Poultry Farming [2007] File No. 220 promulgated by the Ministry of Land and Natural Resources of the People's Republic of China (hereinafter referred to as File No. 220) stipulates that for the land used for the non-poultry farming of the project under the non-local collective economic organization (for example, supporting offices, dormitories, hardening road, etc.), the approval procedure of the conversion of the land for agriculture should be completed and the quota for the local construction land would be taken. However, there had been no quota for the construction land in Pinggu locally. Nevertheless, the File No. 220 also stipulates that the approval procedure is not necessary concerning the poultry farming project created by the rural collective economic organization and the farmers and livestock husbandry economic organization. Therefore, the land could be obtained if the local farmer cooperation partners were introduced.

However, how to organize farmers efficiently and orderly? How to make farmers have motivation to put their land into the project land? Organizing farmers was the strength of the local government. The local government's interest in the project was that the project would not only bring about income to farmers but also drive the small local workshop-styled economy to be upgraded for the industrialization economy, increase GDP, taxes and duties, promote employment, etc. So CP Group and the government of Pingu District reached the agreement of forming a cooperation alliance after brief contact. The government organized farmers rapidly to found "Beijing Lvsefangyuan Farmers' Professional Cooperatives" and it was taken as the initiating subject of the whole project.

Solving the Funding Problem

For an agricultural industrialization project with the budget of over 700 million yuan, how to raise a large sum of money for the subject of the farmers' cooperatives?

In this project, CP Group divided the funds into two parts of the capital and the loan. For the capital, CP Group convinced the government that both parties made 15% of the total investment for the project respectively, and the rest funds 70%, about 407 million yuan could be obtained through loans. However, if the traditional loan guarantee model of the banks was adopted, CP Group could not present so much effective collateral in China and there would no way to obtain the adequate loan for the funds. The financing difficulty aroused in the promotion of the project once again.

CP Group conducted several negotiations with the government of Pinggu District and Beijing Bank. Finally, three parties worked out a new financing guarantee model, namely CP Group and the government of Pinggu District combined their resources that could be used for the guarantee for the bank loan. The resources offered by CP Group were as follows: they pioneered a proposal that they would pay for "Guda" for 20 years without any exception, namely no matter what happened, CP Group management company would pay this sum of the rental fee of the fixed assets and the Lease Management Agreement was taken one important composing part of the combined guarantee in order to proved that the farmers' professional cooperative or "Guda" financing platform was able to repay the capital with interest to the bank. The government of Pinggu District agreed to provide the full amount of discount for the whole process for the loan of the project and most of the money was from various subsidies of supporting and benefiting agriculture.

The reason that Beijing Bank felt interested in the project was that the loan could help it to achieve the target of the loan supporting agriculture every year on one hand and on the other hand this combined guarantee model would not only reduce the risk for the loan but also bring about around 300 million yuan of loan interest to the bank.

Using the Agreement to Bind All Partners

In this model, initiating the project, making the loan and managing the project were to be executed by three parties. Among them, the farmers' cooperatives were the subject to initiate the project, "Guda" company the subject to make the loan and Beijing CP Eggs Co., Ltd. the professional company under CP Group the managing subject. In order to united all subjects into one, CP Group proposed to sign agreements: the farmers' cooperatives and "Guda" signed BOT Agreement, and the former was the owner and authorized the later in the plenipotentiary manner to manage the whole project construction; "Guda" and CP Eggs signed Leasing Management Agreement, and the former leased the completed farm to the later for the management, and the later would pay 12% of lease rentals to the later annually for 20 years. CP Group and the government of Pinggu District signed Strategic Cooperation Agreement and two parties co-founded "Guda" company with the 15% of the total investment as their respective contributions to "Guda" company as its capital. "Guda" and Beijing Bank signed Loan Agreement, and both parties of the joint venture "Guda" provided the combined guarantee for the bank so that it could obtain 70% of the total investment as the loan.

Besides the agreements, the model also stipulated the repayment agreement: CP Eggs should pay 12% of the investment amount to "Guda" as the platform's capital; "Guda" should transfer some of the capital to farmers of the cooperatives to pay for the land rent; the rest money would be the repayment for the loan. For the first eight years, all loans should be repaid to the bank and for the last four years, the capital of CP Group and the government of Pinggu District should be repaid.

Because the output value for one mu was less than 1000 yuan CP Group fixed the land rent at 1000 yuan per mu for farmers in the first year, then it would grow up by 5%. In accordance with such rent calculation, during eight years of the bank loan, each farmer could obtain 5000 yuan for the land rent. During four years of repaying the capitals, each farmer could obtain 7500 yuan. For the remain eight years, each farmer could obtain 23,000 yuan annually.

Such a business model from which all participants benefited was created. CP Group promoted the Pinggu project with 720 million yuan to be implemented successfully with the help of the 6.7 times financial leverage ratio and its self-owned 83.7 million yuan of capital. Through introducing robots, CP Group enhanced intelligentialize and industrialization level of the farming layers in large scale. This fresh eggs production model was of great significance for the whole agriculture production chain development. According to CP Group's views, on one hand it could obtain the fresh egg raw material that CP Group failed to or it was extremely hard for CP Group to obtain by taking other means, which laid the foundation for their subsequent food safety in the industrial line and on the other hand CP Group could also win indirect or extra profit. For example, the planting development such as farming or organic fruits and vegetables in the downstream could bring about additional industrial advantages and profits.

Nevertheless, after CP Eggs agreed to manage this project, it meant that profit should be made that year, since it had to pay 12% of the lease rentals to "Guda" company by the end of the first year.

At the end of 2014, CP Pinggu Layer Farm had been under operation for less than two years. The farm made full use of various means to ensure the fresh eggs' high quality, however, when CP eggs were sold in the market, it still faced lots of competitive brands with slogans claiming that their products had "non-hormone" and "no-pollution". Like Shuxia Sun, the President of Food Nutrition and Safety Committee under China Health Care Society said that "there are no international standard for range eggs and ecological eggs so it's hard to identify their authenticity. In many cases, the sellers just make speculations of concepts to increase prices." One person from Beijing Deqingyuan Agriculture Science and Technology Corporation producing Deqingyuan brand eggs confirmed that "the standards for the ecological eggs are made by the manufacturers and they also fix prices as well."[7]

In such a market where consumers could not find out the true, if CP eggs failed to distinguish themselves and be recognized by consumers for their values, how would their brand values be realized? If the brand valued failed to be realized, how would CP Eggs obtain constant motivation and resources to utilize all machines and supporting resources producing high quality eggs? How would the project of robots farming hens continue to be advanced?

Wang might know better about the interlocking secrets than others and this could be noticed from his extreme attention attached to the market. At lunch one day, when he was told that no CP eggs were seen on sale in one supermarket in the west of Beijing, Wang put down his chopsticks at once and took his mobile phone and called the deputy general manager in charge of sales to check about the matter. At that time, two tempting stewed CP eggs were lying in his plate.

Case Analysis I

Win-win relationships in the "Four-in-One" business model

Shujun Xu

We all know that labour costs in China are increasing. As in most societies with high labour costs, business management in China has to become process-based, standardized and automatic. In this case study, we will look at a scientific approach to hen farming. On the farm in question, robots patrol and monitor certain

Shujun Xu, Associate Professor of the Operations Management Department, School of International Business Administration, Shanghai University of Finance and Economics.

[7]Li [3].

conditions of the henhouse, such as the temperature and humidity, at different times in different locations. They can sense potential problems based on the environment and send devices to solve them if necessary. The indices of the process must be standardised, e.g. scheduled patrolling. Automated hen farming is achieved by sending robots out on regular patrol routes, with detection devices on their heads, chests and knees. There are two major advantages of using robots to raise hens.

First, it can help to reduce labour costs and the possibility of labour disputes in large scale hen farming operations. Using robots has greatly enhanced the productivity of CP Eggs Co., Ltd. since "every henhouse has just one feeding engineer... responsible for managing nearly 170,000 hens". It has also improved supervision as each robot can monitor the environment and comfort of the henhouse 24 h a day (except for charging time). If the operation was fully manned by staff, it would require a significant investment in training, social security and shift work (three shifts) with many potential problems. Although the robots do require a large investment in fixed assets at the early stage, the life cycle cost is greatly reduced. This means more funds can be directed to the "expensive feeding engineer" for a pay rise or training.

Second, it can help to maintain stable product quality. "Hens raised by robots are ensured a stable and comfortable environment", so their egg laying rate and quality is improved. This system can also limit cross-infection between people and hens. In a market where food security is a vital issue, the reliable quality of CP Eggs Co., Ltd. gives them a competitive edge.

Large-scale, specialized and automated hen farming is becoming a global trend. But it does require major investment in many areas and some environmental resources, such as land, can be difficult to acquire. At the same time, the market demand changes quickly and competition is fierce. So creative thinking is needed to solve these problems and to launch new projects.

The essence of the innovative "Four-in-One model" of CP Eggs Co., Ltd. is agility, i.e. to respond rapidly to market demand by integrating all kinds of resources. The CP Group has strong management capacity and the ability to capture market opportunities, the farmers have the land, and the government has strong administration capability. Each group has their own goals: farmers want a stable income; the government is concerned with industrial development, taxation, GDP growth and employment; the CP Group wants to make a profit by implementing the hen farming project and keeping its production stable. Of course, the bank has its own goals (an expected RMB 300 million loan interest) and social requirements (to meet the target for agricultural loans).

In a society where land is collectively owned and the rural population depends on the social environment, companies, governments, banks and farmers all have their own goals. Against this complex background, without creative thinking to integrate resources, launching new projects is no easy task. Considering the investment risk, operational risk and political risk, there is a steep hill to climb. CP Eggs Co., Ltd. have not only adopted customized robots for large-scale hen farming, they have introduced the innovative "Four-in-One" business model to create win-win relationships with the government, the bank and farming organizations, setting a good example for others to follow.

Case Analysis II

The time for robots to replace human labour is approaching fast; innovation is the key

Bofan Zhang

We all know that one of the features of industrial revolution is technological progress. Currently China, like western countries, is going through its fourth industrial revolution, i.e. when robots replace human labour and liberate humans from tedious, repetitive and labour-intensive tasks. The analysis of the robot hen farming case study is timely and thorough. It gives us a clear picture of how to lead the change of industry when the time comes for robots to replace human labour, and offer direction in other areas.

In fact, the analysis shows that replacement doesn't have to mean large-scale lay-offs. Robots can be used for tasks that are highly labour intensive, have safety risks and pollute the environment, and where they can easily replace manual work. The greatest risk for raising hens is the biological security risk. Using robots can secure the health and safety of staff as it prevents cross-infection with the hens. It can also provide a comfortable environment for the hens by keeping the henhouse quiet and reducing the chance of hens being startled by staff entering and leaving the henhouse. This ensures the high quality of the eggs and greatly improves production efficiency.

Currently the biggest difficulty is the major investment in R&D required at the early stage, a significant barrier to entry. Usually, only large companies can afford robots, while medium, small and micro-scale companies with limited funds cannot afford to upgrade.

The innovative business model in this case study is a good approach to cooperation. Using robots to replace human labour can improve the productivity of small and micro-scale companies. The Four-in-One BOT model can help each participant to build on their strengths. The government guides the economy and builds financing platforms with banks to help companies to raise the initial capital for their projects and provide advice on risk control; leading companies develop products and bear risks in their domain; small farms can get more income and enhance their productivity; farmer cooperatives can quickly promote new projects and start

Bofan Zhang, 2012 graduate of the EMBA program of CEIBS, founder of the Baixiong e-housekeeper.

production. By taking advantage of their strong relationship with leading companies, the government can have a positive influence on driving industry development. Leading companies have the funds and capacity to drive industrial robot innovation in their industries and reduce costs. From the bank's perspective, new investment projects alleviate funding pressure on companies and encourage them to replace human labour with robots, as well as expanding their new business models and meeting their targets for supporting agricultural loans. For small and micro-sized companies such as farmer cooperatives, robots are no longer expensive. The BOT provides a holistic low-cost solution, making customized robots and the matching automated devices affordable.

References

1. Han W (2006) "The Large-scale Farming Is the Only Way of Modern Poultry Industry Development", Poultry Society of China Animal Agriculture Association, 2006.11.17, http://www.caaa.cn/show/newsarticle.php?ID=89021. Accessed 11 Mar 2015
2. Zhong K (2013) "President of Deqingyuan Group: Integration Idea of Industrialization Leads Industry Development", China Egg Portal, 2013.01.27, http://www.chinaegg.net/html/n2/4/2013-1-27/201312719524802.shtml. Accessed 12 Mar 2015
3. Li Y (2014) "Labeling 'Title', Price Doubles, Chaos in the Egg Market", people.cn, 2014.01.05. http://shipin.people.com.cn/n/2014/0105/c85914-24025382.html, Accessed 12 Mar 2015

Case V: Alibaba: A Decade-Long Road to Financial Services

In September of 2012 during Alifest, Jack Ma the founder of the world's largest E-commerce platform, Alibaba group, decided the future of the company would focus on three core businesses 'platform, finance, and data'. Since then, Ali has been very active in the field of finance, simultaneously extending new business, launching new products, adjusting their organizational structure and much more. Around this time the slogan "If the bank doesn't change, we will change the bank" stuck a cord that vibrated throughout the world.

However, the finance road for Alibaba was not always a smooth one. With the development of new technologies like big data, cloud computing, platform and mobile Internet and the innovation of commerce pattern like B2B2C and O2O, Ma was not only attacked by traditional financial giants but also was on guard from the variety of challenges brought by Internet upstarts.

With a strong enemies in front and a pursuing forces behind, one has to wonder, "Where is Ma's road?"

This case study starts with 'Trust Pass', that was launched by Alibaba in 2002, and will then tease out the seven key steps that AliFinance went thought during the past ten years. This case study will examine the past, present and future of Alibaba from three perspectives: development path, technological innovation, commercial transformation, challenge and thinking.

This case was prepared by Part-time Case Writer Ji Chendong as the of CEIBS (China Europe International Business School) under the guidance of Professor Zhu Xiaoming for the purpose of class discussion, as opposed to illustrating either effective or ineffective handling of an administrative situation. Ni Yingzi as research assistant of CEIBS made contribution to the case study. The English version of this case study was prepared by Ji Chendong. Certain names and other identifying information may have been changed to maintain confidentiality.

© Springer Nature Singapore Pte Ltd and Shanghai Jiao Tong University Press 2018
X. Zhu, *China's Technology Innovators*, Management for Professionals,
DOI 10.1007/978-981-10-5388-7_5

Table 1 Alibaba's development process

• 1999 Alibaba Group is officially established by 18 founders led by Jack Ma
• 1999 to 2000 It raises $25m from Softbank, Goldman Sachs and Fidelity
• 2002 It becomes profitable, far earlier than many US tech companies
• 2003 Alibaba launches Taobao, the consumer e-commerce website
• 2004 Online payment system Alipay is launched
• 2005 Alibaba Group forms a strategic partnership with Yahoo, taking over the operation of China Yahoo
• 2007 Alibaba.com lists on the Hong Kong Stock Exchange
• 2008 Taobao Mall, now known as Tmall.com, is introduced
• 2011 Alibaba Group announces its plan to build an online of warehouses across China and drive major investment in logistics in the country
• 2012 Yahoo approves a deal to sell about half of its 40 per cent stake back to Alibaba
• 2013 Alibaba names Jack Ma's successor as Jonathan Lu, then executive vice president and an Alibaba veteran of 13 years
• 2014, Alibaba officially started its trading in the New York Stock Exchange
• 2014, Ali Micro Financial Service Group was established, and then renamed as Ant Financial Group

Alibaba

Alibaba group was founded in 1999, by Jack Ma who was an English teacher at Hangzhou Normal University. Alibaba provides extensive services ranging from diversified Internet commerce, including B2B, B2C, payment, and company management software to life classification information.

Alibaba is a global B2B (business to business) E-commerce brand, the largest online trading market, as well as a commercial communication community. Alibaba is headquartered in Hangzhou with 70 overseas branches including in Singapore, India, the UK and the United States. Currently, it has around 20,400 employees.

Alibaba was twice chosen for Harvard Business College MBA case study and has been listed by Forbes, the authoritative financial magazine, as one of the best B2B websites for five consecutive years. Alibaba was also awarded as the most popular B2B website, Chinese excellent commerce website, Chinese top 100 excellent websites, best Chinese trading website and was honored as one of the best Internet commerce sites along with Yahoo, Amazon, eBay, and AOL by domestic and overseas medias as well as foreign venture capitals.

Branch companies include: Alibaba, Taobao, Alipay, Ali software, AliMama, Koubei, Alibaba.com Cloud Computing, Yahoo China, Yitao.com, Taobao Mall, HiChina, Juhuasuan[1] etc. (see Tables 1 and 2, Fig. 1)

---

[1]Alibaba Group Brief Introduction, Chinese Broadcast Network, March 25, 2014.

Table 2 Alibaba IPO Financial Statement Summary

Index	Fiscal year 2013 (2012 Q2–2013 Q1, 100 million RMB)	Fiscal year 2014 (2013 Q2–2014 Q1, 100 million RMB)
Revenue	345.17	525.04
Cost	97.19	133.69
Net profit	86.49	234.03
Ebitda	166.07	307.31
Operating net cash flow	144.76	263.97
Net cash flow from investment activities	5.45	−329.97
Net cash flow from financing activities	−14.06	93.64
Index	2013.03.31 (100 million RMB)	2013.12.31 (100 million RMB)
Cash balance	326.86	436.32
Total assets	637.86	1115.49
Gross liabilities	527.4	707.31
Long-term liabilities	224.62	307.11
Asset-liability ratio (%)	82.7	63.4

Source Alibaba's IPO Prospectus in 2014

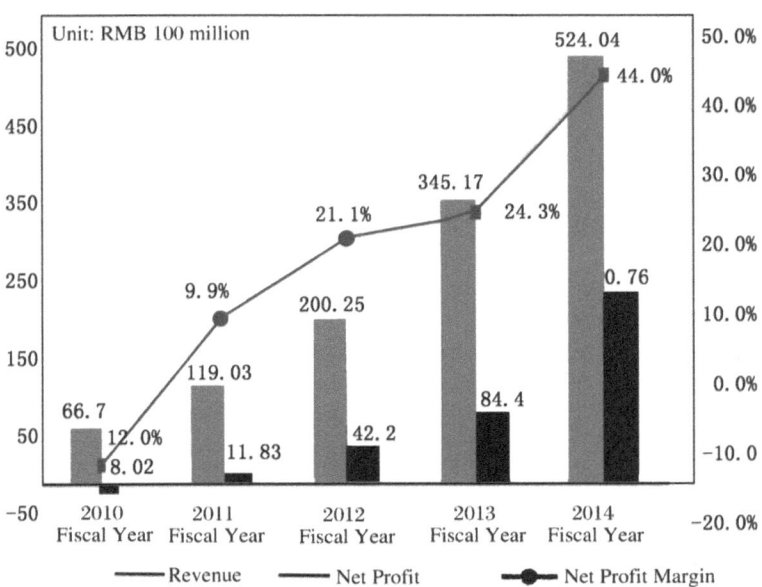

Fig. 1 Revenue and Net Profit of Alibaba (2010–2014). *Source* Alibaba's IPO Prospectus in 2014

Table 3 Three stages of AliFinance's development in ten years

AliFinance's decade layout
Sail with Wind (2002–2010) The first stage
• Launched 'Trust Pass" authentication service to drive establishment of online commerce credit
• Launched Taobao and moving toward hegemony of C2C field
• Launched the third party payment platform—Alipay
• Cooperate with bank to try out online credit service
Diversified Exploration (2010–2012) The second stage
• Established Zhejiang and Chongqing petty loan companies
• Received the first Payment Business License issued by Bank of China. And continued to stretch out into new businesses
• Registered and established guarantee company
• United Tencent, Ping An to build Zhong An Online Property Insurance company
Integrated Layout (2012–2013) The third stage
• Decided platform, finance, and data three business
• Established Alibaba small finance service group based on the four work groups' team and appointed Peng Lei as its CEO
Internet finance creation is for consumers and small enterprises and take platform business as foundation, and data and credit as core

AliFinance

Ten years ago, Alibaba was just founded. In order to finance, Ma who was nursing a dream met more than 40 venture capitalists in one week. But every door he knocked refused to open. In such difficult situation Ma conceived a dream: He wanted to build a platform to help small business grow. This also became the motivation for getting into finance and sowed the first seed for the fast development of AliFinance.

Ten years later, Ma's dream of a finance empire was coming true step by step. From launching 'Trust Pass' to Alipay then to YuEbao, an online credit service that cooperated with banks until in 2013, and an asset securitization program that seemingly came out of nowhere, AliFinance empire began to take shape. The dream of Ma to get into finance was closer and closer to becoming a reality.

An exploration of AliFinance's past ten years can be divided into three stages: Sail with Wind, diversified exploration, comprehensive expansion? (see Table 3).

Sail with Wind (2002–2010)

Ma once said, "Small business' success depends on being shrewd; Medium business' success depends on management; Large business' success depends on integrity." Alibaba is unique in part because credit systems and big data are the key to its financial future. The construction of Alibaba's data and credit system can be traced back to 2002.

Around 2001, the environment of Chinese online E-commerce was in a day to day state of blundering. Some online service vendors were in petty disputes about whether to provide free service or to charge. Then came a unilateral break in the promise of free service and online credit began to shrink.

Given this situation, Alibaba started to rebuild online credit.

In March of 2002, Alibaba creatively launched 'Trust Pass' plan to provide online authentication service for its members. It promoted the establishment of online credit. In the commercial environment of that time, no one imagined that you could build a commercial system on a fictitious online platform with out sacrificing integrity. In the following March of 2004, Alibaba launched 'Trust Pass index' to measure member's credit status with a scientifically evaluated standard.

In May of 2003, Ma launched Taobao.com when he noticed the extensive foreground requirements of C2C (Consumer To Consumer). Ma also launched the third party payment platform Alipay in September of the same year. Finally, in December of the same year, Alipay went live and began to operate independently.

In May of 2007, Alibaba united with China Construction Bank (CCB) and Industrial and Commercial Bank of China (ICBC) and launched online credit and loan products for its member enterprises. After accepting its member's loan application, Alibaba handed over the application and enterprise's credit record, which had been stored in Alibaba's commercial credit data base, to the banks which would then run an independent check and decide if a loan could be issued to the enterprise.

In 2009, Alibaba's online banking department, which was built in order to promote cooperation with banks, carved out from B2B business and put in charge of all of Alibaba's financing business. During the move, the departments were consolidated and renamed as 'AliFinance', so as to complete the transition into being an independent finance business (see Fig. 2).

Above was the first stage of Alibaba's move into finance. Whether Alibaba planned for it or not, it's E-commerce platform and its member' credit data base soon became foundation and its core area of competitiveness. In addition, Alibaba gained a deeper understanding of credit processes and risk control via its

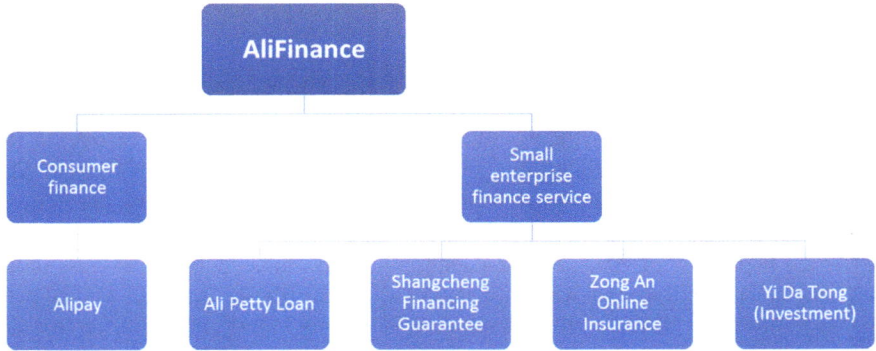

Fig. 2 AliFinance's Business Structure

cooperation with banks. It established a significant foundation for expanding into its next venture: Alibaba petty loan and guarantee business.

Diversified Exploration (2010–2012)

There is no never-ending feast. At the beginning of 2010, after completing an initial try out period, AliFinance and the banks went their separate ways. At that time, Ma had already made preparations for going alone.

In June of 2010, after founding the Zhejiang Alibaba Petty Loan Company, AliFinance credit business was officially launched. In that June, the Chongqing Alibaba Petty Loan Company was also launched. Soon after, Alibaba petty loan companies entered into a period of rapid expansion. Small enterprises were Alibaba petty loan's target and were supported by Alibaba's credit data base and credit evaluation system. The Alibaba petty loan's were non-collateral based, relied on credit to apply for loans online, assured transaction process safety, and promoted user friendly deposit and withdrawal.

At the same time, Alipay expanded across nationally borders without pause and in May of 2011 received the first Payment Business License issued by the People's Bank of China. Meanwhile, Alipay continued to stretch out into new businesses, such as online merchant service platform. They acquired the Anka payment company, got into international airline payment, reinforced their own payment security, and got a fund license of third party payment.

In 2012, Ma' three companies Alibaba, Taobao and Zhejiang RongXin jointly registered and established ShangCheng financing guarantee Ltd. in Chongqing. The company's initial registered capital was RMB 300 million. As a vital part of transaction chain, Alibaba Guarantee Company acted as guarantor providing security in the fields of consuming finance creation and petty loans.

Over three years, Ma consistently researched the fields of commerce credit investigation, the third party payment, online credit, and guarantee so as to insure that AliFinace's diversified exploration went into high gear.

Comprehensive Expansion (2012–2013)

On January 1st of 2013, Alibaba officially begin transforming and clarifying its three main business areas: platform, finance, and data. This period witnessed Ma start an intensive restructuring of Alibaba's organization structure and personnel system centering on these three core areas.

At the beginning of 2013, Alibaba's structure was adjusted to 25 business units but AliFinance and Alipay were not included among these. Instead, Alipay was divided into four business units including a sharing platform work group, an international work group, a domestic business units and AliFinance. Four days later, at the annual meeting for new finance groups, Ma proposed that AliFinance should go back to the essence of finance. He said *"The finance business that Alibaba is doing is not a reform but a revolution, a financial revolution.[2]"* Soon afterwards, Alibaba announced plans to establish Alibaba small finance service group based on the four business units' team and appointed Peng Lei as its CEO.[3]

On October 10th of 2014, Ali Small Finance Service Group was officially established and renamed as Zhejiang Ant Small Finance Service Group. Their businesses included Alipay, Alipay Wallet, YuEbao, ZhaoCaiBao, Ant Petty loan, and was in the process of developing online bank.[4] *"China doesn't need just another finance company but a finance servicing company that focuses on serving small enterprises."* Ma said complacently.

More and more, the overall business arrangement and action order of AliFinance seems to have emerged from the water. Taking Alipay as a fulcrum, the strategic layout of getting into the finance business gradually became clearer. Ali's next ten year focusing strategy started from here.

The AliFinance's Seven Swords Strategy

Jack Ma has repeatedly declared that he won't compete with banks, however, Alibaba will engage in areas that banks choose not to. Ma, a visionary entrepreneur and expert planner, has created many new financial services and products, including credit investigation, payment for goods, absorbing, customer guarantees, insurance, payment and settlement, assets management, etc. Theoretically AliFinance has become a fully functional financial organization. And although, many new financial services are being fine-tuned, it's just a matter of time before they are up and running smoothly.

As is the case in Chinese Kung Fu, Alibaba's every financial move is one step on the path to total mastery. The 'seven swords' strategy integrates different methods all working in concert that step by step consolidate Jack Ma's dream of a financial empire.

Fundamental Credibility

AliFinance based on 'TrustPass', which launched in 2002, laid the foundation for Alibaba's commercial credit system and financial operations.

---

[2]Alipay's 'big ambitions': Chinese Finance 'agitators', China Business News, June 18, 2013.

[3]Note Well: In august of 2014, Alibaba group announced in the new prospectus, according the new agreement of micro finance, that they will share the per tax profits of micro finance as well, not only just with Alipay. Correspondingly, the share of pretax profits that Alibaba group received from last version of agreement were adjusted from Alipay's 49.9% to Micro finance group's 37.5%. In addition, Alibaba group sale small enterprise credit business to Micro Finance Service, the consideration is 321 million RMB plus 7 years annual fee in order to avoid the risk of finance and supervision.

[4]Ali established Ant Micro Finance Service Group, Beijing Times, October 17, 2014.

After three sounds of financing, in March of 2002, Ma launched the TrustPass authentication service on an E-commerce platform. The TrustPass authentication service was Alibaba's solution to solve SME credit issues that occurred during domestic trade. From this, Alibaba's B2B 'membership model' has settled into becoming the company's major source of profit.

Establishing a comprehensive credit check grading system between two transaction parties was an indispensable part of Alibaba's credit investigation process. In March of 2004, Alibaba implemented its 'TrustPass index' in order to bring the identity investigation, history of TrustPass files, transaction status, customer evaluation, commercial dispute and complaints situation etc. of TrustPass members into a unified database. The credit record and evaluation database accurately reflect the production, operation, and sales of online business. TrustPass Index referrals and value are strong due to its ability to evaluate online business' credit status through a scientific quantifying system.

For online business' being a TrustPass member means that an independent third party approved by Alibaba will verify their identity to customers. The online business' transaction histories are recorded, accumulated, and stored in order to win customer trust.

Initially, 'TrustPass' was only designed to tackle online trading problem of asymmetric information, but Alibaba 'accidentally' obtained massive amounts customer credit information. The company put this new information to good use by creating a credit investigation algorithm that resulted in a unique commercial credit validation system for AliFinance.

Payment Rules the Roost

If TrustPass' credit evaluating system and services are only in the intermediary level of finance, AliFinance's Alipay is on a par with a 'grant bank'.

The foundation of all of AliFinance's innovative financial services, Alipay, started earlier and developed faster than any other. There were only three employees in the first Alipay team: a financial manager, an accountant, and a cashier. Presently, Alipay's team has bloomed into an independent payment company with around 3500 employees.

At the end of 2013, users of Alipay numbered close to RMB 300 million with more than RMB 100 million phone users completing RMB 2.78 billion transactions and exceeded RMB 900 billion in payments. This number exceeded the total amount of two mobile payment companies, PayPal and Square, which only managed RMB 300 billion combined. Alipay has become the largest mobile payment company in the world.[5]

---

[5]Alipay official analyzing data, February 8, 2014.

Alipay operation machanism

Alipay is a convenient and safe payment channel, and for bank it is a middleman

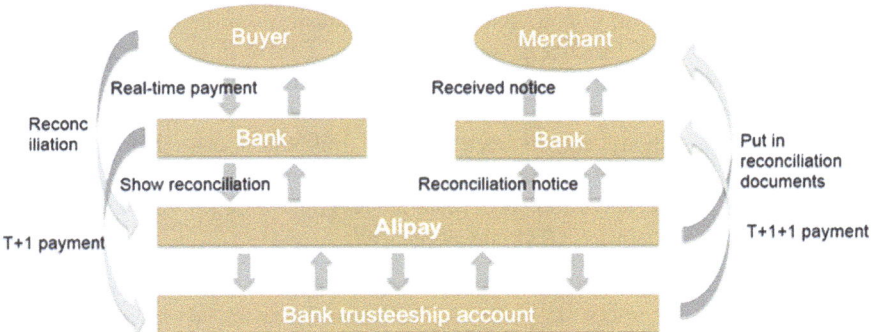

Providing high quality and even customized service or not is the main competitive criteria of payment companies towards differentiation strategy

Fig. 3 Alipay Operation Mechanism

Alipay originally came from the launch of Taobao.com. At the beginning of 2003, after Alibaba's B2B business were making stable profits, Ma went to Japan in order to find a way to further drive up profits. He noticed that Yahoo Japan's localization strategy, stood as the big winner, outdoing eBay Japan, in Japanese C2C market. Meanwhile, Yahoo Japan's CEO and shareholder of Alibaba, Mr. Masayoshi Son (Sun Zhengyi) told Ma, *"Since Yahoo Japan can win Japanese C2C market with localization strategy; there is no reason why Alibaba cannot become successful in China as well."*[6] Traveling through Japan firmed up Ma's determination to launch a C2C business platform. In May of 2013, Taobao went live. The following October, Alibaba launched Alipay, bringing far-reaching reform for finance payment well into the future.

The original purpose to launch Alipay was not to try and solve the problem of online payment, but to build trust between buyers and sellers. On Alipay, buyers can pay, receive delivery notices, and confirm payment after receiving goods[7] (see Fig. 3).

One of the company's most important assets, compared to online payment methods from traditional banks, Alipay's differentiation competitive advantage are as follow:

- Two-way channels: First, Alipay is a fund settlement channel like online banks, which charge merchants in order to make profits. Meanwhile, Alipay is also an exchange channel. Alipay has accumulated about RMB 300 million identified

---

[6]Alibaba Group Business Development History, Snow ball effect.

[7]AliFiance's Previous and Present Life, China Business News, March 29, 2013.

users. Alipay Wallet has exceeded RMB 100 million users. It has become the biggest mobile payment company in the world.[8]

- Online platform to link traditional industries: Alipay is getting into the medical, transportation, catering and entertainment industry via its mobile platform two-dimension code, service window, coupon and WIFI in order to become the preferred online platform of many traditional industries intended to make the move to online Internet.
- Massive data: Alipay accumulated internal transaction data including users' information, consumption habits, amount of consumption, transaction frequency, etc. These users' behavior data and account data provide data support for AliFiance's business R&D and popularity.

In regard to supervision, the Chinese government's attitude toward Internet finance is open but cautious. In particular, the prime minister favorably mentioned Internet finance in his government work report. Internet finance, it was said, should blossom like spring and allotted precious space for Alipay's business.

Aside from the popularity of online payment and third party payment's rapid development, it is the long-term profits born from competition between banks, UnionPay and Alipay that has really caught the public eye. A casualty of this competition was AliFinance's fictional credit card, which went out of business in 2013. In addition, its offline two dimension code and bar code were suspended before launch. Alipay now faced competitive opponents' looking not only to prevent its growth, but were also on attack.

Taking UnionPay as an example, there are three parts of its main income: domestic POS trading income (such as commission from swipe card users), domestic ATM income, as well as international business and explorative business income. Among these, POS trading income accounts for around 60–70% of UnionPay's sales revenue. The use of offline bar code payment and fictional credit card is like expanding payment channels to offline users. When going through a payment channel, the third party payment company, like Alipay, connects directly with banks so that UnionPay was pushed away.[9] Due to this new emerging payment pattern, the market was enlarged which consequently had an absolutely bad impact on UnionPay's marketing status and income. It is no small wonder then that the counterattack of UnionPay and other traditional finance organizations against Alipay is continuing.

Going forward, Ali petty loan company CEO Peng Lei announced her strategy to the public: Wireless and internationalization is the essence of the future. Ali petty loan company's wireless and internationalization will be the key for Alipay's wireless and internationalization.

Alipay's wireless is the core of its vertical market expansion strategy. As early as November 2009, Alipay launched its own mobile payment. With it, users could pay

---

[8]Alipay Become the Biggest Mobile Payment Company in the World, Today Morning Express, February 11, 2014.
[9]Payment War of UnionPay and Alipay, Tiger Sniff Website, March 17, 2014.

money directly to other user's Alipay account, confirm C2C sales, pay water, electricity, and gas bills, as well as put money into their own phone account.

Alipay continuously sought ways to permeate the vertical markets. In May of 2014, Alipay announced a 'Future hospital' plan to the public. According to this plan, Alipay would open platform ability including the account system, mobile platform, payment and finance solving plans, cloud computing, big data base platform, and other services in order to help hospitals build a mobile medical service system. All this was done in the hope of allowing Alibaba users to register with hospitals, await treatment, pay medical bills, get prescriptions, and even interact with doctors and other patients all from their phone.[10] So far, Alipay's business arrangements cover shopping, traveling, airplane tickets, movies, credit payment, public utilities payment, education payment and online games payment, just to mention a few. It is building an 'online life circle' to solve all the traditional problems of clothing, food, accommodation, and transportation for consumers by providing a more convenient personalized method.

The internationalization of Alipay is the core of its horizontal market expansion strategy. Although initial development was rather low key, early 2007, saw the company begin to employ the strategy more and more. By the end of year, Alipay had officially headed out into the international business place. In April of 2014, Rakuten, the biggest E-commerce platform in Japan, began cooperation with Alipay, opening the door for user payment with Chinese Yuan via Alipay on Rakuten. In June, Alipay partnered with American online payment company Stripe who began allowing users to pay with Alipay. Currently, Alipay making moves into the European travel market. Ali members in Europe can use Alipay to get drawback. The drawback will be in their Alipay account as fast as 10 working days and with Alipay Wallet members can check the situation at any time.[11] Alipay's internationalization presence is clear. With Alipay's internationalization, its competitive power will be stronger and market space will be wider.

Petty Loan Gold Mine

Petty loan and micro loan were AliFinance's earliest try out operations; prepared for launch as early as 2007. In 2007, Ali united with CCB and ICBC to provide online joint guarantee loan for membership enterprises. Alibaba provided applicants' membership credit record to banks conducted the risk control and provided credit fund offers.[12]

---

[10]Alipay Launched 'future hospital' O2O to guild an online Medical Dream, Communication Information News, May 6, 2014.

[11]Alipay goes to internationalized and launched oversea drawback service, China Business Website, June 15, 2014.

[12]AliFiance's Previous and Present Life, China Business News, March 29, 2013.

For AliFinance, the impact of this cooperation speaks for itself. First, it attracted more bank dependent enterprises to become a members of 'Trust Pass' and secure the loans. Second, the partnership led to building up a set of credit evaluated systems, a credit data base, and a series of control mechanisms designed to manage with loan risk.

For the banks cooperated, this cooperation enlarged the scale of clients and allowed them to share the advantage of Alibaba's trading platform to help optimize clients' credit evaluation mechanisms as well as perfect loan risk control systems.

However, the three way 'marriage' of Ali, CCB, and ICBC came to an end in 2010.

On June 8th of 2010, with the support and urging of shareholders Alibaba, Yintai, Wanxiang, and Fosun Group, Zhejiang Alibaba petty loan companies were consolidated with a registered capital RMB 600 million. This is the first petty loan company that is fully open to E-commerce small enterprises with a focus toward meeting their financing demands. The new company also obtained the first petty loan company business license in the field of E-commerce. The next year, Chongqing Alibaba Petty Loan Ltd. was established with a registered capital RMB 1 billion. Based on Ali's three platforms, Alibaba petty loan company can also provide order loan and credit loan services. Therefore, Alibaba's finance business 'AliFinance' can be said to be officially started.

As to the percentage of where loans are given, 80% of Ali petty loan goes to Taobao startups on Taobao.com, Tmall.com, and Juhuasuan.com. Regularly, the maximum amount for this kind of loan is RMB one million. The remaining 20% of Alibaba's petty loans are distributed to Alibaba's member enterprises with the regular maximum amount being RMB 3 million (see Fig. 4 and Table 4).

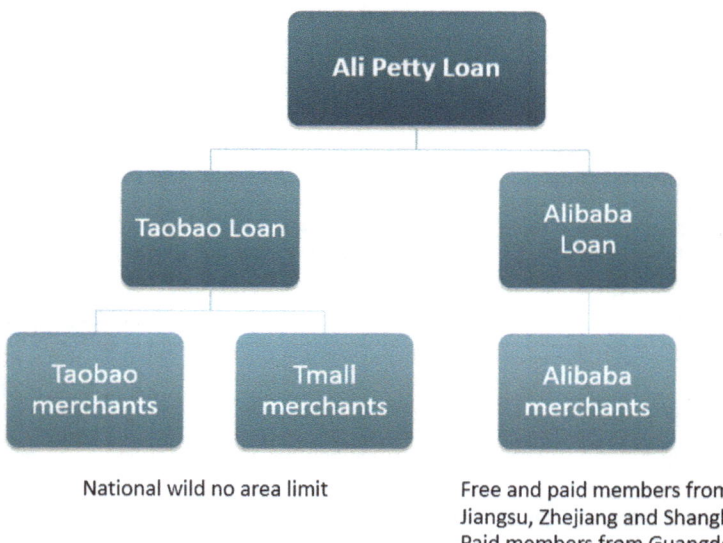

Fig. 4 Ali Petty Loan Business Structure

Table 4 Alibaba Payment Methods

Type	Alibaba credit	Taobao credit	
		Order credit	Trust credit
Credit amount limit	Maximum 30 million CNY	Maximum 10 million CNY	Maximum 10 million CNY
Credit time limit	12 months	60 days	Maximum 12 months
Formula mode	Monthly fixed-payment mortgage	Daily interests	Daily interests
Credit interest rate	Lowest 1.5%/month	0.05%/day	Lowest 0.05%/day
Repayment mode	5 days in advance to notice day of repayment, automatic deduct on Alipay	System automatic repay	Repay interests monthly, repay principle when due Repay fixed interests and principles monthly
Credit process	Fill application online> supplementary information> received credit after approved	Fill application form> confirm page> Apply succeeded (fund flow in Alipay account)	Fill application form> confirm page> Apply succeeded (fund flow in Alipay account)
Application condition	1. Alibaba China member (used to be member of Alibaba's trust pass) or Chinese supplier member, had some operation record (must to be trust pass or Chinese supplier member when receiving credit) 2. Applicant must be legal representative of enterprise or owner of individual business from age 18 to 65 Chinese mainland citizen 3. Industrial and commercial registration were in Shanghai, Zhejiang, Jiangsu, Guangdong and registered time is over 2 years	1. Taobao merchant who is over 18 years old with fully civil capacity 2. Operating Taobao store more than 2 months 3. The owner is honesty and trustworthy and the store has good credit record	1. Taobao merchant who is over 18 years old with fully civil capacity 2. Operating Taobao store more than 2 months 3. The owner is honesty and trustworthy and the store has good credit record

In regard to the cost structure of loan, Ali incurs a 2–5% loan loss provision within 8–18% loan cost structure; this means Ali's revenue is about 1–2.5% of the loan cost.[13]

For small company's side, Ali petty loan has particular advantage in regards loan interest rates and efficiency compared to the traditional banks. First, most of banks' petty loan annual interest rate is around 18% downwards to 13%, but Ali's actual loan interest rate is only 6.7% due to its flexible ways of borrowing and return. Second, Ali petty loan internally has a work pattern called '310', which is, apply within 3 min, award credit within 1 s, 0 employee involved. The whole process is completed online and it is available 24 h 'online'. Having the aid of big data to implement risk control, the efficiency of Ali's petty loan improved rapidly compared to the traditional banks.[14] As of June 2014, Ali petty loan has served more than RMB 0.8 million small companies and awarded more than RMB 210 billion credit loans in total.[15]

Data and Internet are the core of AliFinance' technology. Its natural advantages are its ability to track clients' credit data and behavior that are accumulated on Ali E-commerce platform, introduce network data models, and online video information investigating models. By using cross check technology with the third party verification, AliFinance can confirm the authenticity of user' information. AliFinance is also able to match user's behavior data on the commerce platform to enterprises' and individuals' credit evaluation. Above all, they provide a petty loan option of 'small amount, short period' convenient for those who cannot get loan from traditional finance channels.

AliFinance's micro loan technology underpins its risk control system. AliFinance built a multi-level micro loan risk early warning and managing system. The steps of pre-loan application, in-process loan application, and post loan evaluation are linked one by one. The system utilizes data collection and analysis models, according to small company credit and behavior data to evaluate the enterprise's repayment capability and repayment aspiration. In addition, after loan monitoring and online store/account mechanisms anticipate a user's possible default cost and control loan risk efficiently.[16]

---

[13]Ali petty loan published some core data to cooperate with bank again, Financial Report of the 21st Century, July 23, 2014.

[14]Bank Is Devising Petty Loan Market, PK with Ali Petty loan Who Will Win?, China Securities Website, January 14, 2014.

[15]Yu Shenfa Become General Manager of Ali Petty Loan, Opened Partly Core Data, Financial Report of the 21st Century, June 23, 2014.

[16]AliFiance's Previous and Present Life: Data and Network 'speak', China Business News, March 29, 2013.

Guarantee Assistance

After forming a stable foundation in third party payment and online credit, Ma continued expanding into new financial fields by getting into guarantee and online insurance business.

In September of 2012, Ma's three companies Alibaba, Taobao, and Zhejiang RongXin jointly registered and established Shangcheng financing guarantee company with a registered fund RMB 0.3 billion.

According to an approval document from Chongqing Foreign Trade & Economic Relations Commission, Alibaba occupied 70% of the registered fund, Taobao took 20%, and RongXin the remaining 10%. Company's business scope includes payment of loan guarantee, bill acceptance guarantee, trading financing guarantee, project financing guarantee, LC guarantee, and sideline in litigation preserve guarantee, performance guarantee, and agency service that is related to guarantee business.

Specifically, the leverage ratio of petty loans leverage cannot exceed higher than 1.5 times, however the leverage of guarantee can reach up to 3 times. There is no doubt that entering the guarantee business enlarged company's loan petty limits. Now, the companies can serve merchants who have bigger demands from the fund. At the same time, the guarantee company, acting as guarantor, provides guarantees to consumption finance innovation and petty loan and completes trading processes.

For enterprises, in between the 42 million small companies in China, more than 90% of them need a petty loan. But close to 70% of those small enterprises cannot get loan from the traditional channel because they can't provide a guarantee. Ali Petty Loan provided financing platform and Ali guarantee helped the small business to solve the problems of guarantee from the beginning of company's financing link to the end.

Step Foot in Insurance

On November 6th of 2013, Zhong An online property insurance company LTD became the first that was awarded online insurance license in China. It was registered in Shanghai and the registered fund was RMB 1 billion. In ownership structure, Jack Ma's Alibaba E-commerce company held 19.9%, Tony Ma's Tencent and Mingzhe Ma' Ping An Insurance held 15% separately, the remainder was held separately by YouFu holding, Ctrip.com, and other 3 companies.

From the shareholder structure we can see, Zhong An Online is a pure-blooded Internet company. It also brought a big peculiarity that differentiated it from other traditional insurance companies: 'Three Horses Insurance'. The first, totally automated, online insurance company that is dependent on the Internet to sell insurance policies and settlement of claims; it does not have any branches.[17]

For Ping An, not only was it now able to explore new ways of insurance marketing and solve the problem of channel homogenization, but it also shared the benefits of Ali and Tencent's billions of clients' information.

For the biggest shareholder, Ali, online insurance became a very important fulcrum of AliFinance. Apart from referring using Berkshire Hathaway to obtain a low cost insurance fund by insurance business for commerce reference, Zhong An Online could now assume more important responsibility for AliFinance, that is, learning from methods used before Internet finance, AliFinance followed the good example of using each partner's license, business pattern, and client's resources for their financial products that went to the Internet. Thusly, AliFinance was able to implement the transformation of insurance products and online service patterns to develop the insurance producing chain.

Bao Bao Transformation

YuEbao is Alipay's value added service and the beginning of Alibaba's transformation from trading platform to finance platform. It integrated Alipay and Tian Hong Fund by Ma in June of 2013. The move based on Alipay's platform and since Alipay has the highest frequency of use on the Taobao trading platform, it is very easy for YuEbao to capture massive retail clients.

In essence, YuEbao is a currency fund that uses the Internet for marketing by using embedded direct selling as the main model. From now on, AliFinance can take funds from YuEbao and siphon them off directly to Ali petty loan. All this allows Alibaba to break through the traditional link of supply and demand.

From YuEbao's property combination structure, the bank's deposit and settlement provisions accounted for 92.5%, while bond investment accounted for only 6.7%. That means more than 90% of YuEbao's property was invested in bank's deposit agreement.[18]

For banks, YuEbao inspired finance organizations to see the superiority of E-commerce; but on the other hand, it also made them felt threatened. The increasing of 'Baobaos' including YuEbao steadily began to erode the banks' interests. At the end of January of 2014, the total scale of the online currency fund was close to RMB 1 trillion. Current deposit movement accompanied with a rapid

---

[17]Three Mr. Ma Selling Insurance, Do You Trust Them?, Shanghai Evening Post, February 21, 2013.

[18]Deconstruct YuEbao Ceiling: Survival Pressure of Jumbo Has come, The Economic Observer, January 26, 2014.

increase of liabilities costs magnified the difficulty banks faced with liquidity management.[19]

For consumers, the biggest advantage Ali has compared with the banks is convenience. Although data statistics show that by August 14th of 2014, YuEbao's 7 days of annual yield in June, decreased to 4.19% compared to over 6% in the last year.[20] It remains one of the more attractive financial channels for consumers. To make up for lost profits, YuEbao implemented 'T+0' redemption and fund 'withdraw anytime', allowing YuEbao users to transfer idle money from Alipay to YuEbao in order to buy finance products like funds and thereby make more money. At the same time, users can transfer funds from YuEbao and use the money for some payments function like online shopping or their Alipay account. This 'Catfish effect', forces the banks to innovate their products and services so as to meet the interests of medium and small depositors.

If YuEbao was the beginning of Alibaba's transformation from trading platform to finance platform, then ZhaoCaibao was meant to 'culminating in Alibaba's transformation'.

On August 25th of 2014, AliFinance officially launched an Internet finance product called 'ZhaoCaibao to the public. It was positioned as an investing and financing intermediary information service. It has three main products: fund products, insurance products of all risks levels, and loan products. The anticipated annualized return is between 5.4 and 7.0% with a time limit from 3 months to 3 years.

From direct and indirect financing perspectives, the positioning of YuEbao is very similar to P2P online loan when compared with financing products of banks. Moreover, ZhaoCaibao' supervision department is Shanghai Huangpu finance office district, which is in charge of P2P supervision.

In recent years, the development of P2P has been very fast in China. Statistics show that in 2012 the whole year loan volume was close to RMB 30 billion. However, in 2013 P2P online loan platform transaction was as high as RMB 89.71 billion with year on year growth at 292.4%. It is anticipated that it will maintain 200% blossoming rate for the next two years. At the end of June of 2014, the number of P2P online loan platform had reached 1263.[21]

For investors, the biggest shining point of ZhaoCaibao is 'withdrawing anytime'. Investors only need to pay a transaction fee of around 0.2% of total transaction fee to withdraw money. By not adjusting the income rate of the original product, the problem of 'High income = Long fixed time limit' can be solved. At the same time, ZhaoCaibao enriched the choices of investing products. Not only has ZengLibao which is a currency fund developed by Alibaba and Tian Hong Fund, satisfied

---

[19]Standardize Bank Deposit YuEbao is 'On the List', Shanghai Securities News, February 28, 2014.

[20]Alipay official analyzing data, February 8, 2014.

[21]P2P Network Loan Prosperous future, SOHU Securities, May 28, 2014.

clients' demand for managing money and shopping but so too has the ZhaoCaibao platform satisfied the demand for money portfolio management.

For AliFinance, due to the slow growth of YueEbao, Ali is urgently looking for another efficient model to make up the blank of online financing and trading platform. With the help of ZhaoCaibao, Ali formed an Internet finance system that includes Ali petty loan, ZhaiCaibao, YuEbao, FenQi Gou, and others to constitute a perfect business territory for AliFinance.

Relying on the technology and clients resource advantage of Ali, ZhaoCaibao has a very clear demarcation about its own business. At the core of platform, ZhaoCaibao mainly provides a transaction platform for all sides of financing: While some professional services including relative technology support, data modeling and analyzing, third party payment tools (provided by Alipay), flow supply and project financing are provided by Ali. Thus, such collaboration helps to decrease risks and increase investment income.

First Attempt at a Privately Operated Bank

From 2013, Ali Small & Micro Financial Services Group started the process of uniting with Universal China holding Co. LTD to apply for a private bank license. The project clearly purposed a 'Small deposit, small loan' model and emphasized that the private bank they were committed to would mainly serve small enterprises and community residents based on E-commerce platform. The project has fully taken into account of characteristics of the Internet.

According to CBRC (China Banking Regulatory Commission)'s instruction of private bank pilot project ventilation, Alibaba bank's scheme features were outlined. First, small deposit, small loan model will set the upper limit of both deposit and loan. The feature is clear and conforms to the orientations of different operation. Second, the online bank model by deploying the Internet technology will be used to run bank businesses; E-commerce merchants are their main target clients.[22]

Predictions are that as soon as Ali bank is approved, the lending technique and risk control model of Ali petty loan will be transplanted to Ali bank's 'Small deposit, small loan' model. Ali bank's main business will still be based on big data in order to provide financial service to all enterprises and users of Ali E-commerce platforms.

For AliFinance the temptations of becoming a bank are notable. First, banking is one of the most profitable economic entities in China, it can promote Ali's overall profit rate. Second, banking is the core of finance. The move would further legitimate Ali's financing business. The last, Ali bank will help to raise the overall

---

[22]Ali Exclusively Explain the Absence from First Private Bank, Financial Report of the 21st Century, July 31, 2014.

credibility of Alibaba group and lower the threshold of savings and all kinds of banking business.

However, AliFinance's application to become a private bank has a lot of hitches. On June 25th of 2014, CBRC revealed that they had officially approved the applications of three private banks, including Shenzhen Qian Hai Weizhong Bank, established in Shenzhen, with Tencent and others being main initiators. However, in a surprise twist, Alibaba's private bank hasn't been approved yet.

For the reason for missing the opportunity, Mr. Yu Shenfa vice present of Ali Small & Micro Financial Services Group explained that the preparation scheme of many aspects are still not mature. As long as it is full prepared, they will submit the application to CBRC right away, *"probably very soon, one of the reasons that we will do a pure online bank is because nobody has done it so far. In addition we still need time to do market environment research."*[23]

However, as an important initiative of Ma's finance empire dream, as founding a private bank, Ali still on its way forward.

Traditional Finance

Traditional Finance Operation Model

The most basic function of a financial service to real economy is allocation of funds; that is to transfer funds from saver to investor. There are two types of agencies to match the supply and demand of funds: one is banks, which corresponds to the indirect financing model, and the other is stock and bond markets, which correspond to direct financing. (see Fig. 5) These two types of financing models have a very important impact on resource integration and economic growth but also incur high transaction cost. It directly leads to the profits of banks and securities companies.

However, with the development of big date, cloud computing platform, mobile networks, and finance software, AliFinance on behalf of the Internet finance industry has and continues to deeply change the survival environment and commercial model of the traditional finance industry and draw the attention of traditional finance industry such as banks, securities traders, funds, and finance regulators.

Traditional Finance Industry Is Catching up

Traditional finance has been proactively leveraging its inherent advantages, by uniting with others to dig a deep 'moat' to resist and counterattack the Internet finance industry's capture of territory led by AliFinance.

---

[23]Ali Exclusively Explain the Absence from First Private Bank, Financial Report of the 21st Century, July 31, 2014.

Function

Financing	Pricing	Avoiding risk	Channel
Fast and efficient lead funds flow rationally and improve the efficiency of the allocation of funds	The fluctuant and change price of finance market is a barometer of economic activities	Help to implement risk management, disperse and risk transfer	Reduce searching cost and information cost of transaction

Channel

Regular Bank	Shadow Bank	Private lending (Underground finance)	Capital Market (Direct financing)

Fig. 5 Traditional Finance Organization Business Function

- In September of 2011, Ping An Group strived to build an online financing platform—Lu Jin Suo, making P2P industry a 'Regular army'.
- In June of 2012, CCB (China Construction Bank) launched Shan Rong Business quietly. This is the first one of China's top four banks to launch a comprehensive finance service on an E-commerce platform. Services include B2B shopping mall, B2C shopping mall, and Fang E Tong (a house screening purchasing and individual loan service). It posted nearly RMB 8 billion transaction scale within just one year of operation.
- In the first half of 2013, the online bank business of China Citic Bank (CITIC) launched several businesses including Near Field Communication (NFC), mobile phone QR code payment, and POS online merchant loans.[24]
- In July of 2013, Guangdong Development Bank and E Fund launched 'Smart Gold Account' that was similar to 'YuEbao'. This product was a new credit management tool that combined YuEbao managing money and automatic repayment functions.

Fund companies that were once restricted by traditional sales channels are now fully embracing Internet finance. From building their own online direct sale channels to launching currency cash managing accounts, from T+0 redemption of currency fund and using currency fund payment methods for online consumption. It did not take long time for the traditional finance industry to catch up.

---

[24]China Citic Bank Faster Layout Of Internet Finance POS Online Loan defuse Petty Loan financing difficulties, Hexun Website, January 24, 2014.

Traditional Finance and Internet Finance Industry Win-Win Cooperation

"There are no eternal enemies, only permanent interests." More finance organizations have started to establish win-win cooperation with Internet finance.

Up to now, there are more than 10 fund companies that have set up their 'Taobao flagship store' in the market. In August of 2013, Shanghai Rural Commercial Bank (SRCB) came to a credit payment agreement with AliFinance. Under this model, bank got fund fixed income and AliFinance is responsible for risk control. Franklin Templeton Sealand Fund Management's official website has connected with almost all of the third party payment organization including UnionPay, Alipay, Tenpay, and ChinaPNR etc., in order to diversify its payment channel.

Internet is not intended to overthrow the finance industry, but to change its day to day operations. For consumers, the Internet has bought equality, openness, and sharing. It is driving the Internet finance market competitive landscape to be more optimized and promoting the traditional finance industry to be upgraded and ultimately to assure the maximization of consumers' interests.

AliFinance's Opportunities and Challenges

Alibaba creatively launched 'Trust Pass' member authentication in 2002. The next ten years were spent in a state of constant transformation, always with an eye toward the three core businesses: platform, finance, and data. Ali's clear layout allows them to cultivate in the finance field.

Future Opportunities for AliFinance

The massive customer data base is AliFinance's most core asset. The capability to analyze and apply big data probably ranks it at the top of important boosters for AliFinance's future.

Data Mining Is the Base: It has generated a closed-loop system of data-credit-value in the system of AliFinance. Users' transaction behavior on PC or mobile device generates large amounts of original data that is sent to a distribution center. After passing equipment collection, the original data is distributed to one of a cluster of servers according to internal distribution rules. The original data that was scattered, unordered, and unassociated is processed by the cluster of servers to be readable for people or machines can understand. Data will be mined to form business models. Besides, Ali also built a data exchanging platform in order to solve the internal problems of data's exchange, safety, and compatibility. On this platform, data flows internally to each business group assuring maximum value. Meanwhile, the question of how to provide adequate talents to support professional data mining has become more and more important.

Credit Investigation System is Core: The core assets of AliFinance are the massive merchants and users' credit data base and the credit evaluation systems that were built on Alibaba, Taobao, Tmall, and Alipay and etc. These assets are also coveted by banks and other cooperate partners. These big data and credit systems are at the core of Ali's overall layout for the future of its finance business. Looking ahead, AliFinance will have to prove it is able to play the multiple roles of platform service provider and data supplier in the system as well as solve the problems of consumer's information security. Each of these challenges will test the intelligence of Ali's management team.

Mobile Payment is Strong: While online payment with rapid development of third party payment has been a background focus for AliFinance, their main concern has been the layout of mobile payment technology.

Alipay has successively launched special mobile payment services like bar code cashier, bar code payment, shake payment, two-dimension code scanning payment, 'Yuexiangpai', and sound wave payment based on a combination of its mobile finance payment and commercialization operation experience. It is no doubt that these services have become the standard for mobile finance applications. Ali should have no problem occupying the market as long as the mobile payment market shows an increasing boom, because mobile payment has only just begun to scratch the surface of the marketing space formerly controlled by traditional POS transactions including UnionPay and banks. How Ma will deal with challenges from regulation and the market in the future is worth further observation.

The Challenges Ali Is Facing

Although AliFinance has succeeded everyone inside and outside of the industry is looking askance at it. However, those who don't plan for the future will find trouble at his doorstep. AliFinance is facing successive challenges of supportive platform, growth of petty loan, risk control ability, and sagging YuEbao interest rate, etc.

Platform Business Has Risk of Sagging: with the cold reception to Chinese foreign trade, Alibaba B2B business was affected by the over-all environment and has seen an unstable situation compared to other business. Currently, there are no strong competitors in the C2C field so far. Taobao's flow showed a clear bypass after it launched Tmall. In the B2C field, Tmall encountered powerful opponents— Jingdong Mall, Suning E-commerce, Dangdang, 1st shop and O2O E-merchants. Tmall and Taobao's, the earliest E-commerce platforms, at one time occupied more than 90% of the market, but the rate now has decreased to around 60%.

Three Challenges Alipay is Facing: First, how to deal with boycotts and the impact of UnionPay? UnionPay's main divisions (Union Merchant Services (UMS), and UnionPay Online), have had fierce scrambles with third party payment organizations, including Alipay, in the payment fields of offline POS close order and Internet. Second, Alipay's offline base is weak. Establishing a sales channel is not a short time business. Most offline merchants have built long-term and stable cooperation relationships with UnionPay and their facilities are comparatively perfect. A final question is how to cope with the powerful challenge from WeChat Payment? The most obvious advantage of WeChat is its enormous amount of users. WeChat had a foundation of RMB 438 million active users as of the second quarter of 2014.[25] The feature of social utility gives WeChat Payment a certain advantage in attracting new users and retaining consumer loyality.

Petty loans are restricted by registered capital scale: As a petty loan company, Ali's loan funds to be lent are only limited by its registered capital fund scale. Zhejiang and Chongqing Ali petty loan company's registered funds combined equal about RMB 1.6 billion. According to regulations, petty loan company's financing leverage cannot exceed 1.5 times. The maximum amount of funds the two petty loan companies of AliFinance can provide is only about RMB 2.4 billion. Compared to the enormous client base on the platform, the problem of loan funds source will restrict its development. Moreover, petty loan companies face the cost burden of comparatively heavy taxation.

Risk control model called into Question: The risk control of AliFinance is characterized by the features of background operation and fragmentation. It is not easy for ordinary consumers to understand financial risk. Ali petty loan draws its risk conclusions by analyzing credit level, activity, and the state of operations in each borrower's online store. But if only rely on the credit status of its own platform transaction data, and limited ratio of offline examinations and/or verification, will, in some degree, affect the accuracy of its data and the effectiveness of its risk control. Besides, in recent years a number of corruption scandals involving B2C business were widely reported. Ali faces a great challenge of how to build honest internal control systems, operation mechanisms, and organic safeguards in order to cut the root of commercial corruption.

Interest in YuEbao has decreased with its losing users: Predictably, with the weakness of the currency market, the faster processes of interest rate liberalization, and the launch of bank 'Baobaos', YuEbao, which raked in high profits and was hailed as a water cooler moment earlier, certainly will face a decrease in deposit rate. Users are likely to redeem funds in advance and the fund scale will shrink. With questions such as how not to follow in the footsteps of Money Market Fund (MMF), a PayPal product and pioneer in the third party payment field and how to avoid being squeezed, it is clear the business model of YuEbao will need to address many more fierce challenges.

---

[25]2014 Fiscal Year Second-quarter Earnings, Tencent, August 13, 2014.

Over the past ten years, with the prosperous development of Chinese consumerism and creative implementation of online technology, AliFinance has become a historical commercial legend from the edge of obscurity. Ten years from now, how AliFinance makes Ma's finance empire dream come true will depend on how it deals with the potential opportunities and challenges listed above. Let us wait and see.

Case Analysis I

Knowing Is Hard and Doing Is Even Harder: The Challenging New Circumstances of Ant Financial

Li Yugang

Observing the expansion of Alibaba's financial business, some of its early brands such as cxt.1688.com, Alipay, etc. became significant forces for Alibaba's later participation in the financial services industry, which went beyond Jack Ma's expectations. Henry Mintzberg divides the strategies into two types: deliberate and emergent strategies. Deliberate means that the strategy is planned systematically or designed carefully at the very beginning. An emergent strategy means that it was originally developed to solve specific problems in business operations, but after years of development, it became a key strategic option for the company. The expansion of all Alibaba's financial businesses was the result of both a deliberate and an emergent strategy.

If you analyse the case study carefully, you can see that Alibaba's financial business is characterised by the following elements: First, Alibaba's early financial products were closely linked with its pre-existing core businesses or intentionally designed to complement them. For example, the early positioning of Alipay was to facilitate the development of online trading. Second, each financial product was highly innovative. These products were distinctive in the general market and they were also a brand new type of business in specific segments, meeting the market's specific demands. Third, there were targeted regulations for some new products in the early period, so business operations were on the margins of regulatory compliance. For example, Alipay launched its business in 2004, but it did not get formal approval until 2011. Fourth, the products were highly scalable, which meant they could lead to new business, for example, from Alipay to Yu'e Bao and then to microcredit.

Li Yugang, East China University of Science and Technology, School of Business Vice President, Professor and Ph.D. supervisor.

Since 2015, the Alibaba Group has integrated all of its financial operations into Ant Financial through restructuring. As an affiliated company of Alibaba, Ant Financial has been involved in various financial services, including payment, microcredit, funds, insurance, financial management, credit checking, etc., with brands such as Alipay, Alipay Wallet, Yu'e Bao, Zhao Cai Bao, Ant Credit, Zhima Credit, Mybank, etc. As the business expanded, these financial products which originated from less competitive segments began to threaten the interests of industry incumbents. And with the imitation of latecomers, the company found itself between a rock and a hard place. Alibaba will face huge challenges if it continues to take a traditional approach to business expansion. The following questions need to be addressed.

First, prioritising the expansion of new business while optimising all business types. Taobao helps facilitate Alipay, and then Alipay helps facilitate Yu'e Bao. The transaction data can support credit evaluation, which further promotes the development of microcredit businesses. It therefore seems natural for the company to develop new business and seize market opportunities by leveraging the huge customer resources of Alibaba and Ant Financial. But with rapid and sprawling expansion, decision-makers need to pay close attention to identifying company boundaries, improving headquarters management, not losing focus and avoiding strategic losses.

Second, proper handling of the collaborative yet detached relations between different areas of the business. The success of its early financial services was the result of market positioning and accumulating customer resources. However, with changes in the market, the competition has become increasingly fierce. For example, Alipay now faces competition from many different companies in various ways. As of March 2015, 270 companies have obtained a third-party payment license. So Alipay faces competition not only from the traditional bank card payment business, but also the challenge from Wechat Payment. Ant Financial based its development on the synergy of various business areas, which turned out to an effective strategy, but if one area faces strong competition, then another may be held back. In the early period of a new product, it is reasonable to leverage the resources of existing products and gain support, but in the long run, each business area needs to build independent profitability. For example, Alipay supported the development of Yu'e Bao, but after fierce competition from the overall market, Yu'e Bao no longer has an obvious competitive advantage.

Third, Ant Financial must consider building core competence as an important step in responding to the competitive pressures of the market. Until now, Ant Financial has focused on the "data-credit-value" closed-loop system. As the case study described, "It will be more important to provide sufficient data mining support for Big Data." But as we all know, it will be difficult to establish the relationship between data sets and then provide support for all business areas.

Case Analysis II

Ant Financial: The Disrupter of Traditional Finance

Zhao Zhigang

Objectively, the reason why AliFinance, now called Ant Financial, grew so fast in China was that it was created at the right time. On the one hand, with the rapid development of the Chinese economy, there was a new taste for online shopping and a consumer fashion craze; on the other hand, the general infrastructure such as network bandwidth, online banking interface architecture, etc. were quickly becoming mature. So in this unique situation, Alipay was created as the key closed loop for shopping on Taobao. As the case study described, 10 years ago, Alipay began as a Taobao financial instrument which soon expanded to other e-commerce platforms. The original intention was to solve the credit problems in e-commerce and act as an intermediary between merchants and consumers. Under the conditions at the time, when the Chinese credit system was far from perfect, Alibaba's innovation was bold, resolute and disruptive. The emergence of Alipay strengthened consumer confidence in online shopping and further promoted the boom in e-commerce centered on Taobao. The success of Alipay also set a good example for bold innovation in Alibaba's financial system, including Ali Small Loans, Yu'e Bao, Zhao Cai Bao, YuLe Bao and other subsequent financial ecosystems.

The name, Ant Financial, is particularly good, flipping traditional Chinese ideas about names on their head. Traditionally, words with strong and powerful meanings are favoured, people don't like names that are weak or small. People always like to say they are No. 1 across the country or the world, or even the whole universe, which does not always come true. Jack Ma and his team eventually changed AliFinance into Ant Financial, and fully explained its positioning.

Ants are small but mighty and they live in groups. This name showed that Ant Financial started by serving small businesses or individual members of society, rather than the groups targeted by the traditional banking and financial system. The average transactions were small, not in thousands or millions (you can see in Table 4 that the highest trade amount was 3 million).

Taking Financial combined with Ant, the company aims to provide the ultimate positive experience, making users feel small but mighty. Alipay has been dedicated to serving individuals since its inception, so it has always put user experience first, considering user habits, customer service, payment wait times and other aspects.

So far, Ant Financial has the strongest brand recognition in the Internet industry and this reputation is priceless. I'm sure that in future many companies will follow its lead.

Zhao Zhigang, CEIBS EMBA alumni of Class 2013 and Assistant Vice President of 99Bill.com.

The original positioning of Alibaba was to "create a world where everyone can do business." Peng Lei, the CEO of Ant Financial said during a speech, "We are not a disrupter, but rather a complement to the financial industry. I want to tell the banks and our peers in the financial industry that we don't mean to start a war. There is no need for that at all." It is now obvious that the layout of Ant Financial has evolved into an ecosystem, which is no longer just a credit guarantee and closed transaction loop to meet the expansion demands of Alibaba's e-commerce platforms. Instead, it has attracted a large number of merchants and become an entry controller by building various application scenarios through mobile Internet, Big Data and other new technologies. Its products include Alipay Wallet, Yu'e Bao, Zhao Cai Bao, Ant Credit, Mybank, and data services such as credit checking (Zhima Credit). There are also peripheral ones such as Zong An Online Insurance controlled by Jack Ma, Pony Ma and Ma Mingzhe (insurance business), and the acquired Hundsun Technologies Inc. (providing IT financial system service for public securities and funds). So we can see that the structure of Ant Financial is now more than just a complement. Peng Lei's words were clearly intended to calm his competitors, as the company is still in an intense development stage. Indeed, its low profile reveals its true desire to disrupt the traditional financial industry. Clearly, it would be dangerous to disturb the traditional financial industry in the early stage. They may unite with regulators and overreact to nip the new business in the bud. There is a famous line spoken by Jack Ma in a different context, which is "If the banks don't want to change, then we will change the banks". And sooner or later, that day will come.

As an outstanding example of the Internet age, Ant Financial also faces many growing pains. Pioneers have first-mover advantage, but at the same time, the costs are high. The pioneer needs to coordinate the interests of multiple parties and keep close and frequent communication with the regulators to control entry to their services. And once the model receives approval from regulators, there will be numerous imitators. Another two giant Internet companies, Baidu and Tencent, are also well prepared to develop new services in the online financial industry. In addition, the leading third-party payment companies can also control entry to O2O services by integrating some beneficial offline resources. After all, this is a huge market with vast potential. Considering the nature of an inclusive financial system, a one-horse race will often harm competition and the development of the entire industry. More choices in the market will instead drive participants to provide products that better meet the demands of individual customers, and a smoother user experience and services. We all need to watch this space.

Case VI: Changing with the Times: AutoNavi's Autonomous Development

"From digital maps to automotive navigation and location-based services (LBS) in the times of mobile Internet, I have always been engaged in a sector driven by technology innovation and model innovation."

AutoNavi Software Co., Ltd. ("AutoNavi") Chairman and President Cheng Congwu said.[1]

As one of China's front-runners in mobile Internet services, every move that AutoNavi makes attracts public attention.

On April 11, 2014, the NASDAQ-listed AutoNavi announced that it had entered into a definitive merger agreement to be wholly acquired by Alibaba Group Holding Limited (Alibaba) in a deal valuing the company at about US$1.5 billion. Pursuant to the merger agreement, upon completion of the acquisition, the shareholders of AutoNavi would receive US$21 in cash per American Depositary Share (ADS) of the company. The price represented a premium of 38.5 and 39.8% over the volume-weighted average price of AutoNavi's ADSs during the last 30 and 60 trading days respectively,[2] and was twice as high as the company's record-low share price (below US$10) in December 2012.

Surrendering full ownership to Alibaba, the Chinese Internet titan, is one of the proactive decisions AutoNavi's top management has made for the company's autonomous development.

Since its inception, AutoNavi has been highly sensitive to changing industrial trends, tirelessly pursuing innovation and transformation, and making the right move at the right time: it worked on obtaining certificates of qualification and

---

The case was developed by Professor Zhu Xiaoming and Case Writer Li Yang and Ni Yingzi of China Europe International Business School. The case was developed to provide the basis of classroom discussion rather than to illustrate effective or ineffective handling of a management situation.

---

[1]Zhang [1].
[2]Zhou [2].

© Springer Nature Singapore Pte Ltd and Shanghai Jiao Tong University Press 2018
X. Zhu, *China's Technology Innovators*, Management for Professionals,
DOI 10.1007/978-981-10-5388-7_6

collecting data as a startup, then enjoyed explosive growth, went public overseas, transformed its business, made inroads into the mobile Internet market, marketed its independent platform, adopted a "new mindset" under the weight of intense competition, sold itself halfway through the transformation to join the "big ecosystem", launched its platform strategy in a style of Alibaba...

AutoNavi is both a "model worker" and skilled at grasping opportunities. Its story depicts how an innovative company can "change with the times for autonomous development". Given constant uncertainty and obstacles ahead, how can AutoNavi innovate properly at the proper time?

Starting Up: Arduous Field Work

AutoNavi was born out of an accurate grasp of trends in the automotive sector.

In the 1990s, Cheng Congwu quit his job in the Credit Department of China Science and Technology International Trust and Investment Corporation. The rising commercialization of GPS (Global Positioning System) technology caught his attention, and drove him to explore the potential business opportunities. Inspired by some Japanese companies, he finally set his eyes on automotive GPS navigation systems.

At that time, assured by Chinese government's assertion that "the automotive sector will be one of the central pillars of Chinese economy", as well as by his discovery in Germany that the sales of an auto parts company could be as high as DEM 4 billion, Cheng grew confident in the bright prospects of China's auto parts and accessories industry. "As long as your business is auto-related, it will flourish sooner or later."

Cheng had previously served as President of China Da Tong Industrial Company Limited and Chairman of Huanglong Cave Investment Group. Later on, around 1997, an intuition drove him into the automotive navigation sector, where he engaged professionals for relevant R&D activities. By August 2002, before AutoNavi was officially established, his automotive navigation venture had attracted nearly 10 million yuan in investment.

The founding team of AutoNavi were confident that their instincts would bring them to where they wanted to go. However, they came to realize that it was a road full of thorns later on when AutoNavi was officially launched.

AutoNavi was initially engaged in developing electronic mapping technologies. Electronic maps, or digital maps, enabled by computer technology, provide users with location information that can be stored and accessed digitally. The first batch of digital maps developed by AutoNavi was rejected by its Japanese customer, which left the founding team deeply disappointed but provided an important lesson: to develop a digital mapping business in China, they had to obtain relevant certificates of qualification from the Chinese government first. Unlike Japan and the United States, where the rudimentary mapping databases were open to the public, China exerted strict control over its database. In addition, China's traffic regulations and highway classifications varied greatly by region. It wasn't until 2009 that

China's Ministry of Transport began standardizing the naming and numbering rules of the national expressway network.

"What could you do?" said Cheng. "You had no choice but to start from the arduous field work—collecting geographic information in every corner of the country."[3]

From then on, AutoNavi had two priorities: earning qualifications and collecting data. It was determined to "start with the arduous field work" and build market entry barriers.

In June 2004, AutoNavi became the first private enterprise in China to be granted a Class A Surveying and Mapping Qualification Certificate by the National Administration of Surveying, Mapping and Geoinformation of China (NASMG). In December 2006, AutoNavi obtained the Class A Surveying and Mapping Qualification Certificate for Aerial Photogrammetry; in September 2010, it obtained the Class A Surveying and Mapping Qualification Certificate for Internet Map Services from NASMG.

Qualification certificates alone don't bring market share. Since the digital navigation maps are built upon navigation data, AutoNavi decided to construct its own mapping database through the mass production of navigation data.

The navigation database of AutoNavi is built upon the diligent work of its data collectors, who have left their footprints on nearly every walkable road in cities across the country. In general, the walking data collectors are required to walk 8–10 km a day, identify major buildings, including shop fronts with a width of over 5 meters, in the designated areas, record, check and update the location information of buildings, and take pictures of the overall contours of buildings. The by-vehicle data collectors need to travel 100–300 km per day and often work in a three-person team: one drives the car, another in the passenger seat holds the camera and dictates the road information, and the other double checks and writes down the road information.

AutoNavi's job openings for frontline workers often go like this: "We offer full-time and part-time data collection jobs in cities, towns and villages (day-rate pay)".[4] A typical data collector in Beijing can receive a monthly pay of 3000–4000 yuan and the average age of data collectors is below 26 years old.[5]

By mid-2010, AutoNavi already had a self-developed, comprehensive nationwide digital map database covering approximately 2.8 million kilometers of roadway across China. Additionally, this database contains a searchable and dynamically updated collection of over 12.5 million points of interest, such as office buildings, residential buildings and communities, restaurants, hotels, tourist attractions, gas stations and parking lots. In the course of building its database, AutoNavi has accumulated a massive amount of raw data, including over 364,000 h of video of the roadways across China and aerial images covering over 556,000 km^2.

---

[3]He [3].

[4]Refer to the recruitment information on baixing.com, 58.com and ganji.com.

[5]He [3].

In about a decade, AutoNavi's database coverage has extended from 17 developed cities to more than 360 prefecture-level cities and 2800 counties.[6]

With the qualification certificates obtained and digital mapping data accumulated, AutoNavi defined automotive navigation as its core business.

In 2004, AutoNavi began to provide digital navigation mapping services to Shanghai Volkswagen. In 2005, AutoNavi entered into supply contracts with Alpine, Delphi and Aisin AW to provide digital map data for use in a string of vehicle models, including BMW, Mercedes-Benz, Honda, SAIC General Motors and Audi models.[7]

IPO: A Hidden Champion Appears on the NASDAQ

From 2007 to 2009, AutoNavi's automotive navigation business sustained robust high-speed growth, with revenues increasing from US$21.49 million in 2007 to US$36.21 million in 2009, representing a compound annual growth rate, or CAGR, of 29.8%. Net revenues from the automotive navigation market accounted for 72.5, 64.8 and 63.0% of AutoNavi's total net revenues in 2007, 2008 and 2009 respectively, showcasing its dominant position in AutoNavi's three major revenue sources (i.e., automotive navigation, public sector and enterprise applications, wireless/Internet location-based services).

Qualification certificates and rudimentary data are the core resources of AutoNavi. By leveraging its hard-won qualification certificates and abundant rudimentary data, AutoNavi has built up entry barriers to the automotive navigation market. Just like the German auto parts manufacturer that caught Cheng's attention, AutoNavi also became a hidden champion in the in-dash navigation market.

AutoNavi provides digital map data for in-dash navigation systems installed by over 100 car models, including multiple FAW-Volkswagen, SAIC General Motors, Mercedes-Benz, GAC Honda, Audi and BMW models.

NavInfo is a direct rival to AutoNavi. Unlike the privately owned AutoNavi, NavInfo is hailed as a "national team" in the navigation mapping sector. But for both companies, the in-dash navigation was the major source of revenue. In its annual research report of 2010, NavInfo sounded the alarm, saying "the growth of our major rival AutoNavi has exceeded expectations, squeezing our market share in both the factory-installed navigation and GPS-based mobile navigation segments."[8]

A factory-installed navigation system means that the system is installed as original equipment in the vehicle by the vehicle manufacturer; an aftermarket

---

[6]"AutoNavi CEO Cheng Congwu: A Mapping Life Started from a 'Cowherd' in Jingshan", *Chutian Golden News*, May 16, 2013.

[7]AutoNavi's official website: http://www.autonavi.com/.

[8]Research Institute of China Investment Securities: "NavInfo: Growth Confirmed, Boom to Be Expected", p5w.net, September 13, 2010.

Fig. 1 Sales Forecast of Factory-installed Navigation Systems (*unit* 10,000 sets). *Source* Analysys International, Research Institute of China Investment Securities

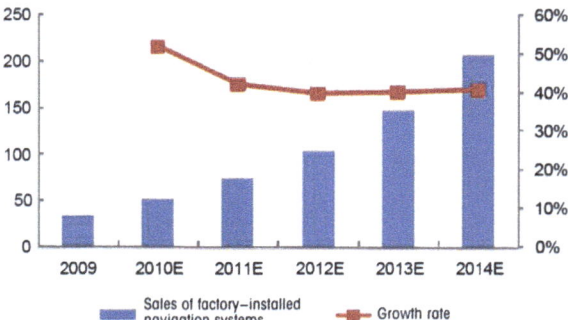

navigation system refers to navigation equipment added to a vehicle as a refitting after it leaves the factory. A factory-installed navigator often outmatches an after-market one in terms of function, stability, ease of use and compatibility[9] (see Fig. 1).

With their respective market share of 48.7 and 46.8% (collectively worth over 95% of the market), AutoNavi and NavInfo became the two oligarchs dominating China's factory-installed navigation market in 2009, creating extremely high barriers to entry (see Fig. 2[10]).

On July 1, 2010, AutoNavi announced its initial public offering of 8.62 million ADSs on the NASDAQ Global Select Market. With the IPO price at US$12.50 per ADS, AutoNavi raised over US$100 million and became the first China-based map data provider listed overseas. By that time, AutoNavi had defined its positioning as "a leading provider of digital map content and navigation and location-based solutions in China".

In the face of stiff competition from the "national team", Cheng stated that AutoNavi would use the funds raised in the IPO to expand its data processing infrastructure, construct R&D centers, and conduct M&A and investment activities in the LBS market so as to enhance its overall strength for sustainable development.[11]

After its IPO, AutoNavi maintained stable revenues from its automotive navigation business (see Fig. 3). Owing to the long-term contracts between AutoNavi and car manufacturers on factory-installed navigation products, its revenues from automotive navigation business were highly predictable over the next three years or a longer period.

Its major rival, NavInfo, went public on the SME board of Shenzhen Stock Exchange in May 2010. In the research report mentioned above, NavInfo

---

[9]*2009 Annual Report of China's Factory-installed Navigation Market*, Enfodesk, Analysys International.

[10]*2009 Annual Report of China's Factory-installed Navigation Market*, Enfodesk, Analysys International.

[11]Huang [4].

The Landscape of Factory–installed Navigation Market, 2009 (market share)

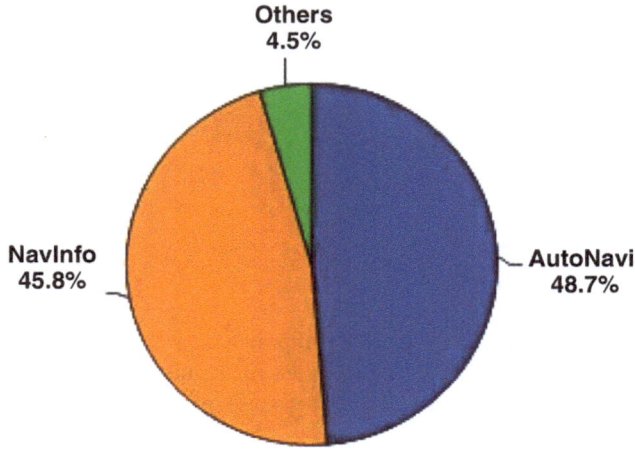

Fig. 2 The Landscape of the Factory-installed Navigation Market, 2009. *Source* Analysys International

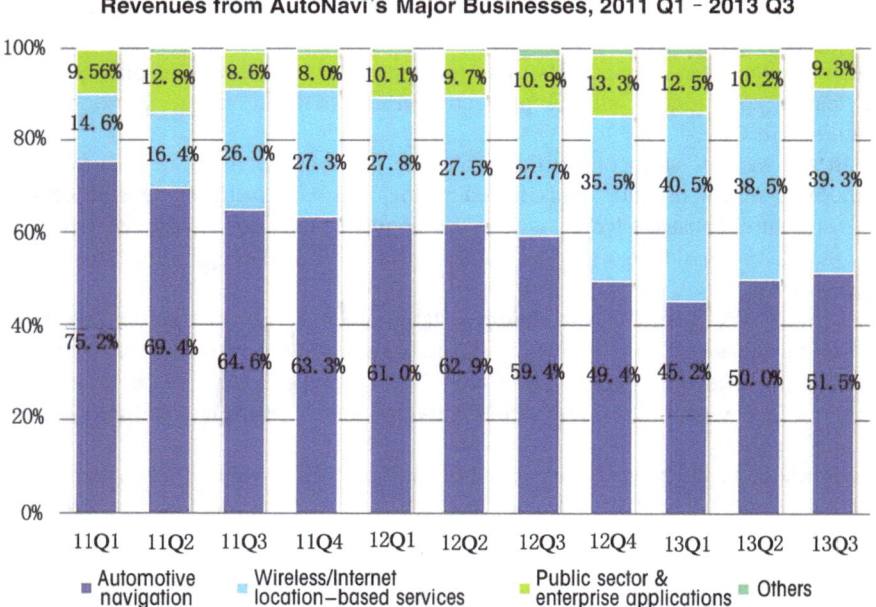

Fig. 3 Revenues from AutoNavi's Major Businesses, 2011 Q1–2013 Q3. *Source* AutoNavi financial reports, adapted by EnfoDesk

forecasted: "Now that we (NavInfo) have begun trading A-shares on the Shenzhen Stock Exchange and AutoNavi has gone public on the NASDAQ, both companies will be further consolidating their market dominance and making their duopoly a longstanding feature of the factory-installed navigation market."[12]

Had mobile Internet never appeared, AutoNavi could have continued relying on its core competitiveness in automotive navigation to keep pace with NavInfo.

Transformation: Embracing the Big Trends in Mobile Internet

After its IPO in 2010, AutoNavi CEO Cheng Congwu keenly observed the fundamental changes brought about by the rising mobile Internet worldwide.

It rose to become a major trend after our IPO in 2010: mobile Internet was gaining popularity across the world. Our map data and navigation products go hand in hand with mobile devices since positioning and navigation services are crucial for mobile devices. Recognizing this major trend, we quickly concluded that the mobile market could mean greater possibilities and opportunities for our company. And hence in 2010, I proposed building AutoNavi into a mobile LBS gateway.

… Mobile Internet would be of great significance to automotive-related businesses. We had to get along well with the mobile Internet trend to consolidate our traditional businesses and seize future opportunities. That was my reasoning at the time.[13]

Cheng reckoned that "the LBS for vehicles are merely a drop in the bucket. In terms of potential user scale, China produces approximately 20 million cars annually, only less than 8% of which have factory-installed navigation systems. In contrast, the number of mobile app users may be in the tens of millions, representing a much bigger market."[14]

The first challenging decision for AutoNavi after its IPO was the mobile Internet-oriented transformation. Pursuing this change when the company was financially comfortable demonstrated the insight and prudence of its top management.

In addition to the top management's insistence on immediate transformation, the decentralized ownership structure of AutoNavi further supported the transformation.

Upon completion of the IPO, AutoNavi's top management collectively held a 52.2% stake, or a controlling ownership of the company. Former Chairman Hou

---

[12]Research Institute of China Investment Securities: "NavInfo: Growth Confirmed, Boom to Be Expected", p5w.net, September 13, 2010.

[13]"AutoNavi CEO Cheng Congwu: Mobile Application as A Strategic Priority", Sina Tech "Global Mobile Internet Conference", http://tech.sina.com.cn/i/2014-05-05/16139359437.shtml, May 5, 2014.

[14]Wang [5].

Table 1 The Pre-IPO and Post-IPO stakes of AutoNavi held by its top management

Top management	Age	Title	Pre-IPO ownership stake (%)	Reduced (%)	Post-IPO stake (%)
Jun Hou	45	Chairman	23.4		19.5
Congwu Cheng	45	Director, CEO	13.5		11.3
Jun Xiao	43	Director, COO	7.9		6.6
Xiyoung Tang	42	VP, Operations	12.3		10.3
Derong Jiang	44	VP, Quality control	5.3		4.4
Nanpeng Shen	42	Director	4.7		3.9
Charlie Yuchen Shi	47	Director	5.6	1.6	3.3
Jeffrey Zhijie Zeng	41	Independent Director			
Dave Qi	46	Independent Director			
Catherine Qin Zhang	44	CFO	2.2		1.8
Yongqi Yang	45	CTO			
Total			64.1		52.2

Source AutoNavi financial reports, imeigu.com, October 29, 2010

Jun and CEO Cheng Congwu became the largest and the second-largest shareholder in the company (see Table 1).

Cheng's dedication to developing the "mobile LBS gateway" implies the company's push to capitalize on its core mapping business to achieve "self-disruption and self-revolution" and create a first-mover advantage in the mobile Internet market.

Take away positioning and you take away all the applications of mobile Internet. … It might have been a little premature for AutoNavi to put forward its mobile Internet strategy in 2010. As the ancient Chinese poem goes, "the duck knows first when the river becomes warm in spring". We could clearly see the signs of the upcoming tide. … AutoNavi has been a pioneer, practitioner and promoter of the industry and has the confidence to take a leading role in the future LBS industry.[15]

To embrace change and new trends meant to redefine the core competitiveness and core resources of AutoNavi. The company boasted ten years of experience in the mapping and automotive navigation sector and had accumulated an enormous amount of data. The vast base of rudimentary data created a huge barrier to entry for

---

[15]"AutoNavi Cheng Congwu: LBS as the Basic Resource of Mobile Internet", qq.com, September 13, 2012.

competitors. As a data-focused company, AutoNavi strived to establish extremely large databases for its major businesses: it won hundreds of millions of digital map content end-users; enjoyed tens of millions of daily active users; received hundreds of millions of search requests; and saw more than a billion searches of its map data. The top management believed that corporate development would be of no strategic significance unless it was built upon the exponential growth of data.[16]

Previously, AutoNavi's strongest products were largely automotive navigation systems and navigation solutions for mobile phones. To respond to the needs of mobile Internet, in early 2011, AutoNavi took the lead and began offering its online mapping and navigation apps for free and providing mobile phone users with LBS (location based service) business dynamics and interactive information.[17] With its proactive approach to mobile Internet, AutoNavi was determined to dive into the blue ocean.

In the digital era, data acquisition is much more than simply copying and pasting existing data. Instead, more data should be collected to enrich its map content and provide users with as much open service information as possible within the user-designated location. This can mean a large amount of work. For example, in the past when the company attached primary importance to its automotive navigation business, the map database was upgraded once a year as required by most auto manufacturers. However, things quickly changed with the rise of mobile Internet, requiring AutoNavi to refresh its database four times a year.[18]

The top management of AutoNavi, including Cheng, believed that in terms of LBS, the data accumulated over the years were where the company's core strength lay, while the data-based technical services (including positioning, navigation, search and map rendering) showcased the company's ability to offer differentiated services. Based on its long-established competitive advantages, AutoNavi was determined to take the first-mover advantage to establish its dominance in the mobile Internet market and position itself as a "gateway to the future of mobile Internet".[19]

When AutoNavi shifted its business focus from automotive navigation to mobile Internet, it encountered internal opposition and pressure from the investment market. Facing these misunderstandings, the top management felt somewhat confused at the outset and unsure of the future orientation of corporate development. The top management maintained their dedication to the transformation strategy.

---

[16]Xiao and Luo [6].

[17]"AutoNavi Cheng Congwu: LBS as the Basic Resource of Mobile Internet", qq.com, September 13, 2012.

[18]Cheng [3].

[19]"AutoNavi CEO Cheng Congwu: Mobile Application as a Strategic Priority", Sina Tech "Global Mobile Internet Conference", http://tech.sina.com.cn/i/2014-05-05/16139359437.shtml, May 5, 2014.

Had they not, it would have been easy to give up halfway, overwhelmed by the opposition voices.[20]

Against such a backdrop, AutoNavi CEO Cheng Congwu reiterated within the company, "A company, either a public or private one, should be able to create long-term strategic value. Otherwise, it won't be attractive to the capital market, strategic investors and potential M&A partners. As long as AutoNavi can produce strategic value in the long run, it will find partners and catch the attention of the capital market sooner or later."[21]

At last, the top management of AutoNavi overrode all objections and brought the company onto a path of reform and transformation. More concretely, AutoNavi defined its mission to offer "mobile location-based service gateway" and vigorously develop the B2C business. AutoNavi's open platform provides third-party partners and third-party applications with an interface with AutoNavi, enabling them to obtain located-based technology and human resource support and services from AutoNavi.

As a listed, specialized company with stable sources of revenues and businesses, AutoNavi seemed to be downplaying its strength with its B2B-B2C transformation. After all, while bypassing its old enemy NavInfo, AutoNavi redefined its target market and business model in a completely different way. Fortunately, the odds were in its favor. The year 2010, when AutoNavi went public, also marked a turning point in China's automotive industry. It was the last year that China's vehicle output and sales sustained annual growth of over 30%. The growth rate dropped to 5 and 10% in 2011 and 2012 respectively. Since the revenue from automotive navigation accounted for more than 70% of AutoNavi's total revenues, the company realized that its core business was reaching a ceiling.[22] The sluggish growth of the B2B market actually reduced the opportunity cost of the company's strategic transformation. In the words of Cheng, "It is our choice whether to sink or swim".[23]

A Self-built Platform: "Penetration into the Whole Industry Chain"

Mobile Internet has brought about changes to the network usage habits of Chinese netizens. As a result, they no longer rely on navigation service apps merely for transportation solutions, but also for a variety of information about everyday life,

---

[20]"AutoNavi CEO Cheng Congwu: Mobile Application as a Strategic Priority", Sina Tech "Global Mobile Internet Conference", http://tech.sina.com.cn/i/2014-05-05/16139359437.shtml, May 5, 2014.

[21]"AutoNavi CEO Cheng Congwu: Mobile Application as a Strategic Priority", Sina Tech "Global Mobile Internet Conference", http://tech.sina.com.cn/i/2014-05-05/16139359437.shtml, May 5, 2014.

[22]Cheng [3].

[23]Cheng [3].

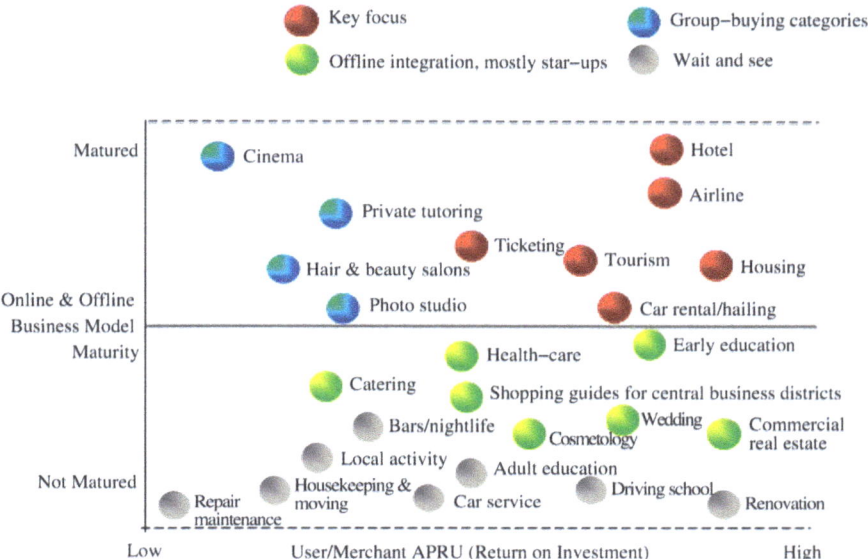

Fig. 4 An O2O Map, by Sector. *Source* Zhang Nan: "AutoNavi's Taobao Dream: Enabling Merchants to Do Business on Maps", Sina Tech, January 25, 2013

such as eating, clothing, housing and entertainment. According to the data of China Internet Network Information Center (CNNIC), of China's mobile phone map users in 2012, 62.7% used route navigation and 45.3% used location search. The proportion of hot spot searches, such as for convenience/lifestyle services, reached 29.2%. For sign-in or location information sharing, this proportion was 10.4%.

The O2O (Online to Offline) e-commerce began to gain momentum. O2O refers to a combination of offline business opportunities with the Internet, positioning the Internet as a front desk for offline transactions. Mobile Internet is closely associated with the O2O model.

AutoNavi Vice President Qie Jianjun believed that maps were the best gateway to all kinds of online-to-offline services. He drafted an O2O map that contained 26 sectors classified by the maturity of the offline/online business model and return on investment. As illustrated on his map, the hotel, airline, ticketing, tourism, real estate, car rental and similar sectors were mature enough and had a high return on investment, and should be key areas of focus for AutoNavi[24] (see Fig. 4).

The open platform strategy of AutoNavi is built upon its core competitiveness and differentiated services. Maps and LBS are supported by its massive data collection, which would be impossible for small and medium-sized enterprises and

---

[24]Zhang [7].

most developers. Fortunately AutoNavi can help developers to develop relevant services and help enterprises to market and distribute their products and services.[25]

AutoNavi redefined its strategy as: building an independent platform through open cooperation, integrating the resources of various innovation-driven enterprises and constructing a LBS ecosystem.

Establishing an independent platform was the second difficult decision made by the top management of AutoNavi, since running an independent platform could mean a great strain on corporate resources. It soon became clear that the decision meant high costs and high risk.

With its open platform strategy, AutoNavi became the technology backer of the location information services provided by a number of Internet applications including Weibo. AutoNavi's positioning as an "LBS gateway" gradually achieved wide recognition in the industry and its penetration into the whole industry chain greatly strengthened the open platform.

However, strategic transformation often involves great financial pressure. As a listed company, AutoNavi's decision to allocate an enormous amount of resources to non-core businesses inevitable impaired its revenues and profitability, worrying capital market investors. Though the future-uncertain reform "moved the investors' cheese", and the large investment in mobile business endangered corporate revenues, Cheng insisted that.

People may feel that, compared with the self-explicit PC-based business model, the mobile world is far more complicated. In my opinion, any mobile application must first survive as parasite in a large ecosystem and strive for a foothold within the ecosystem. Therefore, we should create value for users and gain a sufficient footing in the industry in the mobile Internet times. Only with these conditions are satisfied can we develop our business model in the future.[26]

Standing up to the pressure from all sides, AutoNavi's top management ultimately managed to promote the corporate transformation from a traditional map and navigation service provider to a mobile Internet company, vertically and horizontally integrate the LBS resources, and introduce the open platform strategy to penetrate the whole industry chain (see Fig. 5).

AutoNavi's top management was determined to invest in free offerings out of the belief that a mobile Internet company should focus on building its user base before making a profit, and that user value comes before business value. In the era of mobile Internet, the traditional B2B model, which prized revenue above all else, is not effective anymore. Once AutoNavi got its services right, it would start seeing revenues sooner or later. "Seizing the gateway" was an important move for the company.

---

[25]"AutoNavi Mapping's Open Platform Product Director Fang Fang: Map Platform's Open Practice", CSDN, August 30, 2013.
[26]"AutoNavi CEO Cheng Congwu: Mobile Application as A Strategic Priority", Sina Tech "Global Mobile Internet Conference", http://tech.sina.com.cn/i/2014-05-05/16139359437.shtml, May 5, 2014.

Fig. 5 Industry Chain of Mobile Map and Navigation Market. *Source* AutoNavi's publicity materials. *Note* The industry chain of the mobile map and navigation market can be divided into four layers: ① the application layer, represented by AutoNavi, Careland, Baidu Map, etc.; ② the cloud service layer, represented by Tencent, Baidu, AutoNavi, etc.; ③ the software layer, represented by EMG, Careland, NavInfo, etc.; ④ and the data layer, represented by AutoNavi, EMG, NavInfo, etc. AutoNavi has a presence in all the four layers

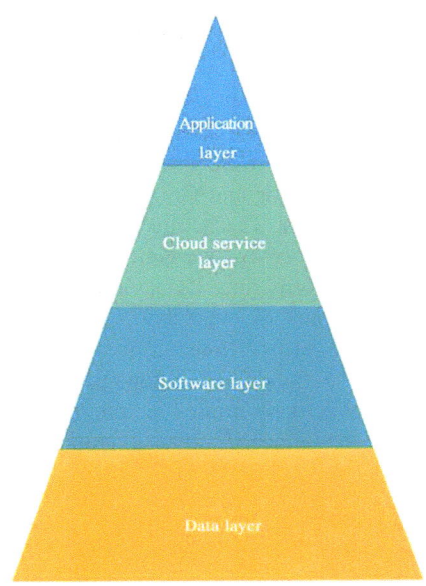

The transformation efforts of AutoNavi received backing from a man with strong views: the Independent Director of AutoNavi and Chairman of Qihoo 360, Zhou Hongyi. Recalling his experience in developing "360 Total Security" at Qihoo, Zhou said approvingly of AutoNavi's efforts, "[AutoNavi] is truly shifting from a traditional software company to an Internet-based company with an Internet mindset and 'online genes.'"[27]

Thanks to its efforts in 2011 and 2012, AutoNavi entered into cooperation with a string of mobile phone manufacturers, including Samsung, Lenovo, Motorola, HTC, ZTE and Huawei, to pre-install its navigation solutions on mobile phones. In this way, AutoNavi seized the lion's share in China's mobile map apps market.

On January 14, 2013, AutoNavi announced that its mobile map app had topped 100 million users in China, and thus became only the sixth mobile Internet company in China with over 100 million users at that time. The other five companies were Tencent, Alibaba, Qihoo 360, Sina and UCWeb (UC Browser). This was an important milestone in AutoNavi's open platform strategy. As of May 2014, AutoNavi's mobile map app had surpassed 300 million downloads.

While AutoNavi was working to penetrate the whole industry chain in the LBS market, the overwhelming mobile Internet tide was driving China's three Internet giants, namely, Baidu, Alibaba and Tencent, to compete for a slice of the LBS market. Throughout AutoNavi's open platform expansion, Tencent and Baidu were never far behind.

---

[27]Zhou [8].

The Demand for a New Mindset

Statistics show that China had accumulated 750 million users of mobile map apps by 2013, of which AutoNavi and Baidu Map accounted for 31.3% (including 6.1% of iOS Apple Map users) and 26.6% respectively, closely followed by Sogou Map,[28] Mapbar and Google Maps.[29] By Q3 of 2013, AutoNavi was ahead of its rivals, with a 32.6% share of China's mobile navigation app market. AutoNavi initially priced its navigation app at 50 yuan and then, when Baidu announced it would be providing its navigation service for free on August 28, 2013, AutoNavi followed suit the next day.

The fight over free services boosted the growth of offline navigation apps, and drove the mobile navigation app market to resemble the map market: the barriers to entry effectively limited the number of later entrants, and market leaders maintained their dominance. Baidu Navigation soon rose to become the No. 4 player in this market.

Though the competition over free services didn't escalate into a subsidy war—as had happened to Tencent and Alibaba when they were marketing their taxi-hailing apps—it was still a money-burning game. In addition to its free navigation service policy, Baidu went a step further in announcing refunds to the previous buyers of Baidu Navigation. This time AutoNavi didn't follow suit. Was AutoNavi financially sound enough to survive a war of attrition against the big-budget Baidu (which had 40 billion yuan in cash, according to its 2013 annual report)?

Despite that some senior executives pointed to "developing AutoNavi into a Taobao for maps",[30] the top management of AutoNavi were actually cultivating new ideas for the open platform strategy. They came to realize that insisting on self-sufficiency across the board could bungle the chance of winning the war, and even mean getting dragged down by bigger rivals. Cheng described the "new mindset" about the platform strategy.

In a mobile Internet environment, any independent application and service market must be a 'parasite' in a large ecosystem. AutoNavi has penetrated every part of the LBS industry chain, but mobile Internet involves a complicated intertwining of small and big ecosystems. Though it provides us with a highly dynamic context and plenty of opportunities, it is still the giants' game. AutoNavi must choose which giant and big ecosystem to side with.[31]

---

[28]In addition to its own map division, Tencent also has a stake in Sogou Map. On September 16, 2013, Tencent acquired 36.5% of Sogou's equity capital on a fully diluted basis for US$ 448 million.

[29]Enfodesk, *Quarterly Research Report on China's Mobile Map App Market, Q3 2013.*

[30]Zhang [7].

[31]"AutoNavi CEO Cheng Congwu: Mobile Application as A Strategic Priority", Sina Tech "Global Mobile Internet Conference", http://tech.sina.com.cn/i/2014-05-05/16139359437.shtml, May 5, 2014.

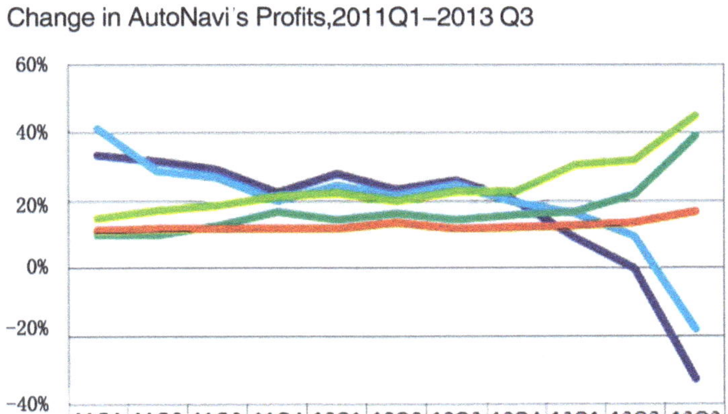

	11Q1	11Q2	11Q3	11Q4	12Q1	12Q2	12Q3	12Q4	13Q1	13Q2	13Q3
Operating Profit Margin	33.5%	31.7%	29.6%	22.5%	28.1%	23.8%	26.2%	20.2%	9.2%	0.0%	-32.5%
Net Profit Margin	41.8%	29.1%	27.3%	20.4%	25.0%	21.9%	25.1%	19.9%	16.6%	9.9%	-17.7%
Percentage of R&D Expenses	15.3%	17.8%	19.2%	21.6%	22.5%	20.5%	23.0%	23.1%	30.8%	32.3%	45.0%
Percentage of Selling Expenses	10.4%	10.3%	13.2%	17.1%	14.7%	16.6%	14.7%	16.3%	17.0%	22.2%	39.4%
Percentage of Administrative Expenses	11.6%	11.9%	12.0%	12.0%	12.2%	14.1%	12.0%	12.5%	12.9%	14.0%	17.3%

Fig. 6 Change in AutoNavi's Profits, 2011 Q1–2013 Q3. *Source* AutoNavi financial reports, adapted by EnfoDesk

In the process of transformation, though AutoNavi had managed to generate stable revenues from its B2B businesses, its total profits had declined sharply due to the soaring costs of R&D and marketing, and of administration to a lesser degree. It became imperative for AutoNavi to decide with whom to side.

In its Q3 Fiscal 2013 Financial Report (as of September 30, 2013), AutoNavi posted total quarterly revenues of US$37.7 million and a net loss of US$6.7 million; its net profits in Q3 of 2012 and Q2 of 2013 were US$10.1 million and US $3.8 million respectively (see Fig. 6).

From a financial perspective, it had been a rough transformation process for AutoNavi, especially taking into account the immense pressure from its shareholders. Reviewing the company's "penetration into the whole industry chain", Cheng pointed out that "it is generally accepted that the future of map business lies in the O2O market. To monetize our business, we need to have a robust business system first."[32] His words implied that it was hard for AutoNavi to continue to maintain self-financing and self-sufficient operations, and that it was time for the company to show its "allegiance" to one of China's three Internet barons—its rival Baidu, or Alibaba, or Tencent.

[32]"AutoNavi CEO Cheng Congwu: Mobile Application as a Strategic Priority", Sina Tech "Global Mobile Internet Conference", http://tech.sina.com.cn/i/2014-05-05/16139359437.shtml, May 5, 2014.

Joining the Big Alibaba Family

Why did AutoNavi accept an acquisition offer from Alibaba? Cheng underscored the importance of "complementarity" between AutoNavi and Alibaba: "Alibaba boasts the resources of tens of millions of merchants, and a powerful payment system, both of which are prerequisites for the commercialization of a mobile app. The combination of Alibaba plus the gateway offered by AutoNavi will enable us to think bigger. We had no reason to say no to Ali's proposal."[33]

In general, it is difficult for companies to generate commercial benefits by providing map services. As disclosed in Baidu's financial statement, the vast majority of its mobile app revenues came from mobile search, 91 Wireless and iQiyi. The revenue from map services was not even mentioned in its financial statement.

"Selling out" was the third of the tough decisions made by AutoNavi's top management. Though selling the company may be the safest way to monetize its assets, the risks of losing independence and weakening incentives are obvious.

The Chinese Internet contains an oligarchy of conglomerates, which ratcheted up the pressure for AutoNavi to transform so it could sustain its competitiveness on one hand, and brought capital and resource integration opportunities to AutoNavi on the other hand. With more than a decade of operations and deployment in the map service sector, AutoNavi had established its unique strategic significance and long-term value.

AutoNavi used to be a company with a dispersed ownership structure and had no controlling shareholder. According to the annual financial statements it submitted to the SEC, as of December 31, 2012, its two major shareholders were Chairman Hou Jun and CEO Cheng Congwu, holding 16.7 and 11% of the corporate ownership respectively.

In May 2013, Alibaba acquired a 28% stake in AutoNavi for US$294 million. AutoNavi was valued at US$1.05 billion in total, and the stake acquired by Alibaba just exceeded the sum of the stake held by Hou and Cheng.

On February 10, 2014, Alibaba announced its plan to buy out AutoNavi for US $21 per ADS in cash—a total of US$1.045 billion. Upon completion of the deal, AutoNavi would become a wholly owned subsidiary of Alibaba and be included in Alibaba's ecosystem. In nine short months, Alibaba increased its valuation of AutoNavi to US$1.45 billion.

In total Alibaba invested US$1.339 billion to acquire AutoNavi's 253 million mobile app users, 300,000 developers, map database, well-developed mapping teams and technical teams, as well as its close ties with car manufacturers.[34] All previous shareholders of AutoNavi, including Cheng, sold their stakes in the company.

---

[33]"AutoNavi CEO Cheng Congwu: Mobile Application as a Strategic Priority", Sina Tech "Global Mobile Internet Conference", http://tech.sina.com.cn/i/2014-05-05/16139359437.shtml, May 5, 2014.

[34]"The AutoNavi Map Effect", *Business Review*, April 8, 2014.

After the acquisition deal, it was time for Alibaba to integrate its e-commerce database and AutoNavi's database, and to combine its merchant relationships, cloud computing and payment strengths with AutoNavi's positioning, navigation and map search abilities. Backed by its big data resources, the new version of AutoNavi Map unveiled in April 2014 successfully incorporates the Alipay function, provides a variety of information about everyday life (covering catering, hotels, entertainment, real estate, etc.), and is becoming a daily life service gateway connecting users, merchants and developers.[35]

From its mobile Internet-oriented transformation into a self-built platform, pitched battles against the titans and final sale to Alibaba, the development of AutoNavi represents an irony: AutoNavi was a "good student" who always had the right answer at the right time, but it was still pushed by the "invisible hand" of fate into an acquisition.

Platform Strategy in Alibaba's Style

AutoNavi has keenly closely watched and embraced market trends. Starting from its navigation service business, AutoNavi executed a series of transformation decisions to establish a first-mover advantage in the mobile Internet market. After jumping into the blue ocean under immense competitive pressure, AutoNavi rose to be a new darling of the capital market and won unique opportunities for development.

Despite good prospects, the post-acquisition integration has been arduous. In the first nine months after Alibaba became its shareholder, AutoNavi failed to post expected growth in market share even with access to Ali's merchant data and payment platform. Its biggest rival, Baidu Map, in contrast, steadily expanded its market share from 24.4 to 26.6%.[36]

AutoNavi's old enemy, NavInfo, had obstinately clung to the automotive navigation market. Facing the overwhelming mobile Internet tide, NavInfo finally announced in May 2014 that it had accepted Tencent's offer to buy an 11.28% stake of NavInfo for 1.173 billion yuan. Upon completion of the deal, Tencent would become the second largest shareholder of NavInfo.[37] As a result, Baidu, Alibaba's AutoNavi and Tencent (NavInfo) came into closer quarters with each other as they competed in the navigation/map/LBS markets.

Since being acquired by Alibaba, AutoNavi has had two lines of business to focus on: one is the B2B Internet of vehicles, i.e., cooperating with car

---

[35]"AutoNavi Launches Offline Navigation—A Shift towards 'Gateway' to Daily Life Services", ifeng.com, April 11, 2014.

[36]"Five Challenges Facing Alibaba after Its Acquisition of AutoNavi", Sohu IT, February 12, 2014.

[37]"Tencent's Investment in NavInfo Divides Map Market into Two Camps", qq.com, May 6, 2014.

manufacturers and providing them with location-based interaction services; the other is the AutoNavi Map and AutoNavi Navigation B2C business.[38]

AutoNavi made its fortune from automotive navigation, and the increased investment in mobile Internet businesses doesn't mean neglecting the automotive market. As shown in the research report released by iiMedia Research in 2014, the brand awareness of AutoNavi and NavInfo in the factory-installed navigation market was 92.1 and 75.3% respectively.

Since the Internet of vehicles is becoming a new source of business opportunities for the integration of the traditional industry chain, AutoNavi is well positioned to "launch a surprise attack". In the meantime, more and more car manufacturers have begun taking the Internet of vehicles seriously. Qoros Automotive is one such example. Its Internet of vehicles platform offers over 30 innovative services, as well as access to AutoNavi and Sina Weibo. More specifically, the in-dash touch panel integrates a wide spectrum of functions, from microblogging and sign-into navigation, car status monitoring and restaurant booking.[39]

Compared with the mobile map market, the Internet of vehicles represents a blue ocean, primarily home to Internet titans, cloud platform operators, car manufacturers, telecommunication operators and automotive navigation/O2O service providers. It is still a game of giants. In this context, AutoNavi can bring the "parenting" advantage into full play: it can act as a content integrator to help car manufacturers draw in various content resources, leverage the transaction tools available to get through the closed loop of operation services, and provide navigation and payment services at the same time.[40] Having been acquired by Alibaba, AutoNavi is expected to do a better job in this aspect.

AutoNavi's new boss has unveiled a blueprint for mobile Internet. In June 2014, Alibaba fully bought out UCWeb and formed the Alibaba UC Mobile Business Group. The valuation of UCWeb was estimated to be up to US$5 billion, and Yu Yongfu, Chairman of UCWeb, would lead the new Business Group. After the merger, the new Business Group would oversee the LBS operation, i.e., the B2C business of the former AutoNavi.

After joining Alibaba, the stress AutoNavi shouldered has been greatly relieved and its drive to seek out opportunities gradually attenuated. With the support from Alibaba, the transitioning AutoNavi no longer has to rush to make money. In September 2014, Yu Yongfu announced that the new AutoNavi would focus on user needs and the R&D of mapping and navigation technologies, and would not be required to reach any specific commercial objective in the next three years. "We have the perfect conditions for product development. Making money is not the top priority. Instead, we are busy considering how to spend money. And users are our

---

[38]"Renaissance of AutoNavi: Yu Yongfu Take the Reins, Refocus on Map Business", DoNews, September 24, 2014.

[39]"Qoros Auto Builds Leading In-dash Information & Entertainment System", mycar168.com, May 23, 2013.

[40]"The Times for AutoNavi Automotive Navigation", China.com, April 24, 2014.

primary focus. After the privatization, the Internet business team of AutoNavi breathed a sigh of relief."[41]

The most explicit change is that AutoNavi began to purify its product lines by eliminating those functions that had been commercialized prematurely, while also avoiding group-buying and other similar products that may impair the user experience.

Compared with the previous self-built platform, Yu Yongfu believes that the new platform strategy is more in the "style of Alibaba": "We hope to establish clear accountability and shared responsibility, and promote win-win cooperation with the developers—AutoNavi provides the LBS expertise, and the developers focus on application service innovation. A healthy LBS ecosystem can thus be established in the form of AMAP Inside."

According to Yu, AutoNavi will open all user services based on LBS to its partners and encourage entrepreneurs to take advantage of "AMAP Inside" to build up a robust LBS ecosystem.[42]

Yu is still furthering the strategy of "LBS rather than O2O". The industry believes it was a wise decision for AutoNavi to put aside the O2O business since it is still too early to march into the O2O market with map products; adequate development of the LBS business will drive O2O consumption sooner or later.

The integration is underway, and challenges are emerging in an endless stream: how can the conflicts of interest among different business units within Alibaba Group be alleviated? How can maintain its ability to see key business trends? What can be done to keep the AutoNavi teams motivated?

Initiating a mobile Internet–oriented transformation, building an independent platform, joining the big ecosystem … Can AutoNavi continue marching to the drumbeat of innovation and reform—or will it miss a step?

In the constantly changing world, innovation and reform are a marathon.

Case Analysis I

A Tough Entrepreneurial Path: Innovation and Difficult Choices in AutoNavi's Development

Yu Xiubao

This is an excellent case study of entrepreneurship and innovation. Presented in chronological order, it gives a full picture of how AutoNavi was set up, developed

Yu Xiubao holds a Ph.D. degree from University of Queensland and works as Associate Professor at School of Economics and Management and Director of DBA Program in Tongji University.

---

[41]Yu [9].
[42]Yu [10].

and driven by innovation against the backdrop of fast economic growth in China, and how the Chair Cheng Congwu led his team to apply innovative approaches and make difficult choices about the Company's future.

The case study discusses the growth and transformation of AutoNavi in 7 parts. The first part (Starting Up: Arduous Field Work) describes how Mr. Cheng grasped the start-up opportunity and collected geographic information from every corner of the country to develop the company's navigation database. The second part (IPO: A Hidden Champion Appears on the NASDAQ) is about AutoNavi's business growth during China's economic boom and especially the peak of the auto industry, as well as its performance after going public on NASDAQ. The third part (Transformation: Embracing Major Mobile Internet Trends) is about the tough choices for business transformation (tapping into mobile Internet) as mobile Internet went mainstream. The fourth part (A Self-built Platform: "Penetrating the Entire Industry Chain") discusses how AutoNavi focused its business on people's everyday lives. Parts five (The Demand for a New Mindset) to seven (Joining the Big Alibaba Family; Platform Strategy—Alibaba Style) discuss the decisions made by senior management, facing cut-throat competition as the industry matured.

This case study covers the following key points:

Firstly, having an innovative product and business model is of vital importance. Innovation was the cornerstone of AutoNavi's survival and development. During the early stage, the Company developed a product called "electronic map navigation" for in-dash navigation systems. In 2010, it carried out R&D for mobile Internet products, releasing an "online map navigation mobile app", followed by a "location service portal", offering user-friendly Internet services. These products and service models were the result of huge R&D investment and manpower input and helped AutoNavi gain the upper hand in industry competition.

Secondly, startups need to build competitive advantage in the early phase of development. Porter's Competitive Advantage Theory says that each company needs a unique competitive advantage in order to survive. AutoNavi owned the largest database of underlying map data in China, which was both a valuable resource for the company and a guarantee of high quality service for customers. This reliable database served as a solid foundation for AutoNavi's market-oriented moves after 2010.

Thirdly, it is vital to identify opportunities to develop a new business, apply innovation and make bold decisions. In this case study, we see how NaviAuto's founder Cheng Congwu made important decisions at different stages of China's economic development. (1) When the government made clear in early 2000 that the automotive sector would be a central pillar of the Chinese economy, Mr. Cheng sensed an opportunity and quickly decided to start up his business; (2) 2007–2009 witnessed the rapid growth of AutoNavi, followed by its listing on NASDAQ, providing a capital foundation for later development. (3) Riding the wave of mobile Internet, the Company launched a free mobile navigation service in 2011; (4) As mobile Internet developed further, AutoNavi began to penetrate the entire industry chain, offering services that were relevant to people's everyday lives. When making critical decisions about the future, corporate leaders cannot be 100% certain of success. This case study shows us the background to their decision-making and the struggles of senior management facing difficult choices.

Fourthly, starting up a new business is hard work. The first part of the case study describes the arduous field work, collecting geographical information for the Company's mapping database in the early stage. This seemingly unconventional approach involving a lot of manual labour was an important way for AutoNavi to build its competitive advantage.

Case Analysis II

From Blue Ocean to Red Ocean: How AutoNavi Shifted its Goal from Money-Making to Money-Seeking

Li Yuan

The AutoNavi case study describes a typical struggle for innovation and business transformation in a company.

In its early days, AutoNavi accurately spotted the vast potential of the electronic map market and car navigation market, as these products were rarely found in China. Starting from scratch, AutoNavi made a bold leap ahead of the competition through data collection work to provide rich data resources. When both business and individual clients gained disruptive and unprecedented value from its service, the market was a blue ocean for AutoNavi. The value curve commonly used in a blue ocean strategy could even be ignored. That is because AutoNavi's services were unprecedented, from a unique customer experience to meticulous data collection, triggering strong user demand. Their products changed people's habit of following a map to being guide by voice navigation. AutoNavi also targeted the factory-installed navigation sector as its primary goal. Though NavInfo was a strong rival, the market was big enough for the companies to divide up and still make a handsome profit. The scarcity of mapping certificates and the investment required for data collection were effective barriers to the entry of new challengers.

Subsequently, this market did become increasingly complex. Luckily, Cheng Congwu identified the opportunities brought by mobile Internet at just the right time. He said, "Take away positioning and you take away all the applications of mobile Internet." By virtue of its data advantage and differentiated services for navigation and searching, AutoNavi turned itself into a "mobile Internet entry point", diving into a new blue ocean.

Li Yuan is an alumnus of CEIBS EMBA 2011 and the Asia-Pacific Director of Electronic Security at Delphi.

Below is a comparison of three value curves.

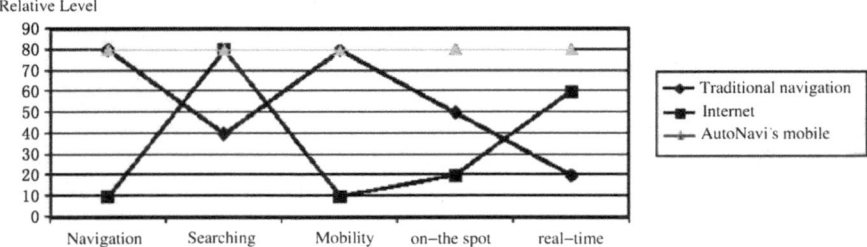

- AutoNavi's value curve is obviously different from traditional navigation and the Internet.
- On each of the key indicators, AutoNavi performs at least as well as or better than traditional navigation or the Internet.
- Distinctive cross-sector products meet market demands.
- A blue ocean takes shape!

The barriers to new entrants were still very high, such as developing a database, maintaining a data advantage and maintaining the existing customer base. At first glance, this seemed to be a blue ocean, but that was the wrong conclusion. AutoNavi ran into a common problem for start-ups: it's easier to start a business than it is to keep it going as the blue ocean suddenly turns into a red ocean.

How can a market with high barriers and obvious differentiation be a red ocean? Perhaps we need to see it from a different perspective. Instead of analysing potential market entrants, we need to consider whose market segment was under threat. AutoNavi wanted to enter the mobile Internet sector, where business models were broadly the same throughout the Internet industry. This meant that AutoNavi's efforts threatened the interests of the BAT Internet giants (Baidu, Alibaba and Tencent). Crossing the barriers of mobile Internet was a piece of cake for these corporations. When Baidu entered this sector, it climbed quickly to No. 4 in a short time. So in fact, AutoNavi had not discovered a blue ocean, instead it had jumped into the red ocean dominated by large corporations. We can also see from the case study that AutoNavi invested heavily in both its O2O and LBS businesses, but didn't earn enough revenue to make a profit (see Exhibit 6 and 7).

There's always a gap between a beautiful dream and reality. Fortunately, AutoNavi's dream was attractive enough to sell for a good price to an Internet giant. If the original goal of AutoNavi's chairman and CEO was to sell the company, then they successfully achieved that. In fact, we can detect a stronger desire to "create value" than to "make money" from Cheng Congwu's statement about corporate strategy ("A company, either a public or private one, must be able to create long-term strategic value. Otherwise, it won't be attractive to the capital markets, strategic investors and potential M&A partners. As long as AutoNavi can produce strategic value in the long run, it will find partners and catch the attention of the capital markets sooner or later.").

AutoNavi's adoption of a mobile Internet strategy has some parallels with the automobile industry where I work. For instance, an entrepreneur wants to produce spare parts from a new material which is sturdier, cheaper and lighter than existing materials, which has never been used before. The market potential is huge, millions of cars would benefit from it every year and the company would have first-mover advantage when it jumped into this blue ocean. But in fact, the entrepreneur would be taking market share from spare part giants. Spare parts made of the new material would encroach on the market of traditional spare parts, forcing industry giants to fight back. They could easily crush any entrepreneurs with their strong R&D resources, procurement and bargaining power, business relationships and economies of scale, turning the blue ocean into a red one where only big corporations can survive. In these circumstances, selling at a good price is perhaps the best option for a startup.

After AutoNavi was acquired by Alibaba, the main concern shifted from making money to user experience. It's now part of a bigger game. AutoNavi has always been good at attracting customers, collecting data and delivering service, and these abilities have been integrated into the Alibaba empire. It now plays a different role in business operations. Rather than rushing to make money and generating profits, AutoNavi concentrates on providing location-related services such as the current LBS and possibly O2O services in future. It would become an entry point for users, a collection point for user data, a delivery point for the Alibaba service, and ultimately the detector of unique opportunities for Alibaba. It's no exaggeration to say that AutoNavis acts like an unconventional marketing tool for Alibaba. Compared with traditional B2C marketing approaches such as advertising and promotion, AutoNavi offers real-time and on-the-spot marketing that's more targeted and effective and more compatible with Alibaba's system.

In conclusion, AutoNavi took the initiative to dive into blue ocean by entering the navigation market. However, in its subsequent innovation process, it made no distinction between blue ocean and red ocean products, which made its acquisition by Alibaba inevitable. The lesson to be learned from this case study is that whether you are devoted to growing your business, or you want to sell it in a short time at a high premium, you must explicitly define your objectives for innovation and startup. You should also make a clear distinction between blue ocean and red ocean markets, and list the intended results you hope to see in the expected period of time, in order to formulate the corresponding strategies.

References

1. Zhangni. AutoNavi Chairman & CEO Cheng Congwu: Exploring new profit model of the navigation sector. Global Times, 2 Dec 2013
2. Zhou J. AutoNavi & Alibaba Entering into Merger Agreement. Shanghai Business Daily, 14 Apr 2014
3. He Y. Cheng Congwu: AutoNavi is an inevitable choice for everyone. Entrepreneur, Dec 2012

4. Huang S. AutoNavi President Cheng Congwu: Bigger investment in R&D, More M&A activities. cnsoftnews.com, 31 Aug 2010
5. Wang W. AutoNavi & self-directed development. Business Value, 23 Oct 2012
6. Xiao Y, Luo W. The mobile mapping world—a dialogue with Cheng Congwu. IT Times Weekly, 5 Apr 2013
7. Zhang N. AutoNavi's Taobao Dream: Enabling merchants to do business on maps. Sina Tech, 25 Jan 2013
8. Zhou H. AutoNavi's transformation from software company to internet company. xinhuanet. cn, 30 Aug 2013
9. Yu Y. AutoNavi to elevate from software company to internet company. China Business News, 24 Sept 2014
10. Yu Yongfu to Build AutoNavi Open LBS Platform, 'AMAP Inside' rather than O2O. National Business Daily, 4 Nov 2014

Case VII: ICBC in the Digital Times

Internet Financial Counter-Attack

In the past more than ten years, the disintermediation wave struck the whole society. Especially when some non-financial enterprises used the mobile internet technology to infiltrate into the financial field rapidly, the internet finance was born. This new business gave birth to the new environment and pattern of the non-bank payment and financing intermediary so that the traditional banking was challenged seriously. In such environment, how should we make adjustment? How should we follow the trend?

On June 29, 2014, standing on the podium in "Master Class" of China Europe International Business School, Mr. Jiang Jianqing, Chairman of Industrial and Commercial Bank of China (hereinafter referred to as "ICBC") asked such questions and opened the curtain of his strategic thinking about ICBC in the digital times.

"The internet finance's challenge against the commercial bank does not lie in technology but in thinking and the way of thinking. You may know the internet very well, but without the internet thinking, you will be behind the times". Mr. Jiang Jianqing said, *"Therefore, if we want to rebuild up a totally new ICBC, we should not only make use of the internet technology but also introduce the internet thinking"*.

"However," speaking of which, he could not help but frown, *"ICBC's traditional mode of thinking and financial inertia is very powerful."* Mr. Jiang Jianqing pointed out that this was the greatest obstacle to rebuild ICBC.

So, could Jiang Jianqing break this obstacle?

The case was co-written by Professor Zhu Xiaoming, Case Writer Zhu Qiong and Research Assistant Ni Yingzi of CEIBS. During the writing, they also obtain cooperation and support from ICBC. The case's purpose is to be used as the material for the class discussion and it doesn't indicate whether the case mentioned here is effective or not.

X. Zhu, *China's Technology Innovators*, Management for Professionals,
DOI 10.1007/978-981-10-5388-7_7

ICBC Informatization

Jiang Jianqing and Informatization

"*It is most difficult to change human's mode of thinking.*" In order to prove his view, Jiang Jianqing cited an example that he experienced himself. Around 1982, People's Bank of China introduced computers to do accounting instead of hand-work, and this could have been a thing to improve efficiency, however, employees did not trust computer. They used abacus to re-accounting for checking results calculated by computer. When two results were not consistent, employees thought that it was the computer that made mistakes. However after several times of accounting, they found out that it was the human who made mistakes. Employees had used this method to check computer's accounting for several years, and finally they came to believe that the computer makes no mistake indeed. "*The computer is used to replace the handwork accounting. The traditional thinking inertia was so powerful, let alone what we are facing today is much greater and deeper changes in thinking!*"

"*So does ICBC have no hope to realize the changes in thinking?*" Hearing this question, Mr. Jiang Jianqing made a brief meditation and said slowly, "*I think that the human being's knowledge will change slowly in the constant practice and different environment. A small number of people will change firstly and gradually more and more people will change. Therefore, we can make use of the information technology to create such process and environment.*" Mr. Jiang Jianqing pinned hope on his familiar information technology.

In July, 1999, Mr. Jiang Jianqing left the position of Branch President of ICBC Shanghai for Beijing. He worked as the Deputy Party Committee Secretary and Deputy President of ICBC in Beijing. In February, 2000, Mr. Jiang Jianqing served as Party Committee Secretary and President of ICBC. His book "*Study on Financial High-tech Development and Its In-depth Impact*" on the relations between financial development and high technology was completed in 1999 and was published in 2000. In this book, Mr. Jiang Jianqing wrote: "*The history of the whole banking financial business development is a process in which the financial business is innovated constantly and the high banking technology develops and applied continuously. In fact it is the development history of the banking electronization.*"[1] Therefore, "the future competition in the banking sector is more and more a competition of banking science and technology innovation and a competition of information technology."[2]

Mr. Jiang Jianqing's deep study of information technology started in 1995. When he went to America for further study, he coincided with emerge of American Security First Network Bank. This is the first bank in the world to offer the internet financial services. It was open online in October, 1995. For a few months after it

---

[1]Jiang [1].
[2]Jiang [1, p. 89].

was open, nearly 10 million person/time browsed it online, which brought about a great shock to the financial sector. At that time Mr. Jiang Jianqing was in the shock front accidentally, he read a large number of relevant media reports. The internet finance concept was rooted in his brain at that time. "Then it was not like now that people were able to collect electronic literature with computer conveniently. Many materials were copied by me at the time." With those materials, he wrote main chapters of another book during his stay in America—*Technological Revolution in American Banking Industry.*

With such research accumulation, after he became the executive of ICBC Head Office, he immediately listed the information technology as his important work. Therefore, "9991 Project"—a large data concentration project which was two years earlier than the domestic counterpart was launched by ICBC in September 1, 1999. A few years later, an experienced person in the technology department in China Construction Bank commented that "ICBC was the first in China to propose the implementation of the data concentration, and it implemented the strategy very decisively. Today facts prove that ICBC's decision is of strategic significance."[3] "9991 Project" lays the leading foundation in the information technology for ICBC in later days.[4] Since then, ICBC's informatization entered into the rapid development orbit. From 2005, ICBC invested 10 billion yuan in the informatization construction annually, concerning building various types of technological hardware and services, technological human resources and others.

When the bank began to face the internet finance challenge, Mr. Jiang Jianqing said, "*Whether in thinking or in the financial innovation, we are well prepared.*" Qian Bin, General Manager of ICBC Data Center (Shanghai) partially verified Mr. Jiang Jianqing's preparation. "*To make the big data analysis needs long-term plan and accumulation. Without our chairman's early requirement in the term and our constant data accumulation for years, we won't have foundation even if we want to make the big data analysis today.*"

With well-done preparation, when Mr. Jiang Jianqing discussed the general innovation framework of the bank development with professional departments of ICBC. Mr. Jiang spoke clearly and logically so that his subordinates dared not have slightest neglect. "*When they see me, they feel dizzy. Because I not only tell them strategies but also a large number of tactics, and even discuss in detail.*" Speaking of this, Mr. Jiang Jianqing felt proud in face in some way. Most of business executives in ICBC proved that their development strategy even some tactic details were from the chairman initially.

Mr. Jiang Jianqing's requirement for the innovation task made his subordinates feel so stressed that they were almost breathless. At a meal, the person in charge of the Electronic Bank Department reported to him excitedly, "*Our Wechat Business volume reached 290,000 transactions per day.*" But Mr. Jiang Jianqing replied with

---

[3]IMB "Rebuild Nerve Center—ICBC's Road to Large Data Centralization", June 13, 2010, http://www.cbinews.com/casestudy/news/2803.html.
[4]Wu [2].

"too little. Our online banking (users) has reached nearly 180 million, and a batch conversion should be made so that the Wechat Business can reach at least the level of above 100 million." When the ICBC e-commerce platform—E-Purchase had been online for four months, Mr. Jiang Jianqing said, *"Our big data analysis is not done enough, and I think a large number of quality data analysis reports should be made out of the platform. This is the reflection of lacking the internet thinking."*

This helmsman in ICBC had always been using this way to "whip" ICBC mercilessly to run forward.

Informatization Process

Information System

ICBC was founded on January 1, 1984. As of the mid-2008, its profitability level jumped to be the first in the global banking industry and then it was in the leading position all the way. As of the end of 2013, ICBC's total asset was 18.917752 trillion yuan with a year-on-year increase of 7.8%, the total debt 17.639289 trillion yuan with a year-on-year increase of 7.5%, the profit 262.965 billion yuan with a year-on-year increase of 10.2%. With this achievement, ICBC ranked the first in 1000 large banks in British Bankers, and also the first in 2000 global enterprises in American Forbes and became an important global bank.

It was ICBC's globalization business network that made such achievement. By the end of 2013, ICBC's business had stepped across six continents globally and extended the overseas network to 40 countries and regions. Through 17,245 domestic institutions, 329 overseas institutions, 1903 agents, online banks, mobile banks and self-service banking, ICBC provided financial products and service for 4.735 million corporate customers and 432 million personal customers.

What made this global network play competitiveness was ICBC's informatization operation management system. After 30 years of informatization construction, ICBC business operation management had gone through three times of evolution and formed four systems (see Table 1). The first generation (1995–1998) system was the core business system with the account as its center. On the basis of the traditional counter, the business operation turned into computerization from handwork accounting gradually; the second generation (1999–2002) CB2000 system was the comprehensive business system, realizing the unification of account accounting, fund remittance and transfer and clearing and 7X24 business services; the third generation (2003–2008) NOVA system was the data concentration processing system, concentrating customer information, realizing the multiple channel management and proving the informatization management means for the risk management and decision-making, the fourth generation (2009 to now) NOVA + system was to support the internalization and integration business development further on

Table 1 ICBC informatization operation management system evolution

System	Core Business System	CB2000 System	NOVA System	NOVA+ System
	Customer-oriented	Comprehensive teller	Centralized data processing	Internationalization, Integration
System	Core Business System	CB2000 System	NOVA System	NOVA+ System
	Customer-oriented	Comprehensive teller	Centralized data processing	Internationalization, Integration
Features	From handwork to automation Support the traditional counter services	Unified financial accounting Capital remittance settlement 7x24 hour business services	Payment through multiple channels Customer information centralization Operation management Risk management Decision-making support	Relatively independent accounting Rapidly innovated products Management information centralization Comprehensive risk management Reasonably level-division and Flexible and loosely coupled application system
Stage	The First Generation	The Second Generation	The Third Generation	The Fourth Generation
	1985-1998	1999-2002	2003-2008	2009 to 2014

1999-2001

In Sept. 1999, the great centralization of data was launched officially.

Solve 2,000 computer problems.

In October, 2000, ICBC The corporate sub-system of Comprehensive Business System (CB2000) was put into operation successfully

2002

The 2nd phase of CB2000 was put into operation successfully.

2003-2005

In Dec. 18, 2003, the personal online banking brand Finance@Home was launched.

All-function banking system (NOVA V 1.0-2.0) was put into operation successfully in the whole sector.

2006

The core application restructured project was put into operation successfully.

The promotion of the personal credit management system (PCM2003) was completed.

2007-2008

The overseas comprehensive business processing system (FOVA) was put into operation.

The online banking restructured project was put into operation in the whole sector successfully.

"1031" project construction was launched.

2009

The 4th generation application system NOVA+1.0.0 version was put into operation.

3G mobile banking was launched.

2010

Eight platforms were constructed successfully.

The promotion of FOVA system was accelerated.

2011-2013

The "12th Five-Year" application plan was launched.

The central services were integrated taking responsibility for production and maintenance.

Source: ICBC

the basis of the overall framework improvement, and achieved separation of accounting from products, comprehensive risk management and management information concentration.

Independent R&D

Almost all banks didn't neglect the informatization construction. Why did only ICBC deduce such a system evolution and achieve the global competitiveness? *"Just depend our strength to conduct R&D"*, Qian Bin blurted out. Qian Bin fully understood the answer to such question. Because he noticed other banks' embarrassment and even nothing gained when those banks depended on external strength and he was involved in reaping ICBC's self R&D achievement.

Before 1996, ICBC's system was purchased from outside. On June 29, 1996, ICBC established Software R&D Center in Zhuhai. Since then, ICBC entered the independent R&D times. The center was composed of 150 employees in the beginning, who were all from various branches of ICBC and familiar with the banking business. Meanwhile they knew something about the technical framework platform. Those engineers were called to Zhuhai but their personal relations were still remained in their former units. They worked in Zhuhai with 1200 yuan per month as subsidy.

The team completed two large projects in the beginning of 1999 after its development. First, it established ICBC capital settlement system so that the time for each exchange difference transaction was shorted to 2 h from 24 h; Second, in Shenzhen Branch, the first core business processing system was online as pilot. *"These two systems development lays foundation for the subsequent large data centralization,"* commented Qian Bin.

"ICBC was not the first bank in China to set up the software development center among its peers", Qian Bin disclosed. At the early 1990s, another bank in China established the R&D center not only in Paris and Tokyo but also in Shenzhen, but the road it took was different from ICBC's. The office building was purchased and researchers were employed from other companies. After several years, no product was developed and finally "the building was deserted".

However, ICBC's independent research and development aroused more demands of business departments when products were launched. So Zhuhai R&D Center was broken into four R&D centers. Guangzhou center was established in October, 2002, taking responsibility for the business system research and development such as personal finance, credit card, overseas and technical platform; Shanghai R&D Center was established in October, 2003, taking responsibility for research and development of human resources, financial management, business management analysis, risk management and office management; Beijing R&D Center was established in May, 2004, taking responsibility for online banking, telephone banking, mobile banking and overseas research and development; Hangzhou R&D Center was established in December, 2006, taking responsibility for research and development of asset management, financial market, intermediary business and non-bank finance.

Data Center

While ICBC Software Development Center had certain foundation in 1999, The "9991 Project", an unprecedented data concentration project in China's banking history, was launched. Why was the great data centralization implemented? Lin Xiaoxuan, chief information officer in ICBC presented that "the great data concentration is an inevitable choice in the commercial bank development. Without the great data concentration, we could not achieve the management concentration, risk control, product management and decision-making support and others were out of the question."

The "9991" Project took three years. The double-center mode design program was adopted. 37 ICBC computing centers were concentrated to Beijing and Shanghai. The core business system with unified process standardization was realized under operation 24 h a day and 365 days a year.

On October 27, 2002, Shanghai branch was merged to Data Center (Shanghai), and Beijing branch was merged to Data Center (Beijing), marking the successful end of the project.

In September, 2004, ICBC conducted the physical transfer of Data Center (Beijing) and Data Center (Shanghai) and clarified their functions: Shanghai center was the production center in ICBC and Beijing center was the backup center for production center and business test center. In 2014, ICBC Shanghai Jiading Data Center was founded and ICBC realized two-region and three-center operation and maintenance mode. Shanghai Waigaoqiao and Jiading data centers realize double-center operation in the same city, independent deployment, failover and short-distance and offsite disaster backup while Beijing center was used as the disaster backup for these two centers.

ICBC data centers served not only internal ICBC but also the third party cooperation partner or other enterprises seeking trusteeship such as trusting data of fund companies or insurance companies, cooperating with them to launch products, even using technical strength of ICBC data centers for developing management systems and maintaining them for in-depth cooperation partners like China Exim Bank. "Of course, we also determine a very strict limitation to assure security of cooperation partners' all information," said Qian Bin.

On the basis of the physical integration of data centers in 2004, ICBC implemented restructuring of core application systems in Data Center (Shanghai) in 2006, and the restructuring result was that the corporate banking and personal banking business were separated logically in application system. Zhang Ying, deputy general manager of information and technology department in ICBC, said that "*such separation promotes flexibility. The company and personal backstage data are running respectively and they don't influence each other. When relevant products are researched and developed, research and development will speed up because of mutual independence.*" Furthermore, Zhang Ying believed that the separation meant the bud of the bank's big data way of thinking. A deputy general manager of ICBC Data Center (Shanghai) gave a further explanation that when the separation actions were taken, ICBC was also taking the big data issue into account. "*With the business scale increasing constantly, the backstage system will reach the extremely big data level. With the separation of corporate and personal businesses, we don't need to face too large amount of data when we carry out subsequent horizontal expansion or tap data.*" However he added "*we have not defined that ICBC's big data started from that time.*"

Business Electronization

"*Business electronization is one of decisive factors for ICBC's subsequent development*", said Qian Bin.

ICBC business electronization started in 1996. By the end of 1999, 44 large and medium computer centers were established, the electronization outlets reached 28,000, 3400 ATMs and more than 20,000 POS terminals. The first level network with the head office center connecting 43 first-level branch centers, the second level network with the first-level branch centers connecting all subordinate prefecture and municipal centers and the third level network with all prefecture and municipal centers connecting all subordinate institutions were established. Intermediate business such corporate, savings and various acting receipt and payments were electronized and businesses such as ATM, telephone banking, electronic exchange and Peony Card were launched.

In 2000, ICBC founded the e-banking department. From 2000 to 2003, ICBC clarified its development strategy with the online banking as focus, telephone banking as popularity and mobile banking as exploring channel. In June, 2001, ICBC online banking (e-bank 3.0) was online, providing the corporate online banking business firstly. Then the platform successively added other traditional businesses in ICBC such as personal banking business, bank card business, private investment and other business. In December 8, 2003, ICBC launched the personal online banking service with "Banking@Home" as a brand. Since then, ICBC online business has covered all corporate and personal business.

From 2004 to 2008, ICBC improved its e-banking competitiveness through building up "four channels" and "six platforms", realized it's "expanding" market strategy in the online banking, in which, "four channels" refers to online banking, telephone banking, mobile banking and self-service banking; "six platforms" refers to capital management, e-commerce payment, charges payment, financial investment, acting sales and marketing service platform.

After 2009, ICBC made the development strategy, leading the domestic e-banking business development and creating the first class e-bank. Since 2004, ICBC international settlement system, trading financing, acting credit management, foreign exchange centralized unchanged, forward foreign exchange and other systems were put into operation successively.

With the constant development of the overseas business in ICBC, as of 2013, all businesses in ICBC were processed through computers. E-banking transactions with online banking as focus accounted for 81% of all businesses with the transaction volume more than 350 trillion yuan. The online banking users reached 170 million and the mobile banking and telephone banking users broke through 100 million. The strength of the online banking in ICBC also reflected in another group of data. In accordance with the released industry analysis data, in the second quarter of 2013, in the rankings of the transaction scale of the online banking market in China, ICBC ranked the first with 38.01%, the next two were China Construction Bank (16.33%) and Agricultural Bank of China (12.06%).[5]

Such data reflected the independent development result of informatization in ICBC for 20 years. Up to the year of 2014, ICBC had 13,000 technicians, including 120 employees in the technical department of the head office, 4300 employees in the software R&D center, 1100 employees in Data Center (Beijing), 780 employees in Data Center (Shanghai) and technological team with 7000 employees in branches.

"Having such information technical capability, why do we worry about the internet finance's counter-attack?" Standing in the podium, asked Mr. Jiang Jianqing.

Counter-Attack of Internet Finance

Internet Finance

American Security First Network Bank was under operation officially in 1995 and it was purchased by Royal Bank of Canada in 1998. The bank was regarded as the early bud in the internet finance. Nevertheless, another two forces were active in the market. One was internet enterprises. With the e-commerce or other internet channels, they grasped the customer's cash flow and information flow so that they stretched to the customer's payment and financing and other financial fields. Another was financial enterprises, which made use of internet tools and extend to customers' other financial or non-financial fields from payment financing. Obviously, two powers were heading toward each other from two opposite directions.

To the financial enterprises, Hou Benqi, General Manager of Electronic Bank Department in ICBC, said "Using the constantly developing IT technology to develop business is a systematic process. However, internet enterprises that emerge unexpectedly spoil our pace. We achieve the same from different roads with them, so we become competitors from cooperators.

The first internet enterprise involved in the financial field was Alibaba. In 2012, in Alibaba network business conference, Ma Yun, Chairman of the Board said that Alibaba would be divided into three main businesses: platform, finance and data. "It is not because we want to make more money but we feel that in the times, we need to use the internet thinking and internet technology to support reconstruction of the

---

[5]Enfodesk, 2nd Quarter of 2013 China Online Banking Market Industry Database, July 13, 2013, http://data.eguan.cn/qitashuju_170458.html.

future financial system in the whole society." "In the next five years, Alibaba will apply for its own bank," one executive in Alibaba disclosed.[6]

Alibaba's financial mode got the support of Zhou Xiaochuan, President of People's Bank of China. He said that competition (introducing the internet finance) improved the traditional industry development that made it adapt to new condition with a sense of irritation so that it helped them keep up with the pace of the times and technology.[7]

With Zhou Xiaochuan's support and example of Alibaba, very soon Tencent, Jingdong, Baidu and NetEase entered this field in different ways successively. From 2013, the internet finance craze had been spreading in the domestic market.

Mr. Ma Weihua, the former President of China Merchants Bank, thought that in addition to encouragement and support by the regulatory authority, the objective reason for the internet finance craze was the market space arising from insufficient services for vulnerable groups offered by traditional financial institutions, such as the small and micro enterprises loans, personal loan guarantee, microfinance, P2P and others. Serving those "long tail" markets was precisely the strength of internet enterprises. They made use of "Big Data, Cloud Computing Platform and Mobile Internet" technology to process massive amount of data exactly, efficiently and economically and found out valuable decision-making information so that the financial innovation services became possible. Of course, with rapid increase of netizens, the online consumption and life had become a common practice. This was also an reason that should not be neglected.[8]

There were three major business features in internet enterprises' involvement in finance: (see Table 2):

• Through adoption of new technologies, customer information was grasped and credit risks were managed;
• On the basis of the point-to-point transaction, the financial resources were allocated;
• Capital transfer on the third party payment.

With those businesses, the internet enterprises ran between financial enterprises and final customers. If they gained a foothold in these businesses, the financial enterprises would be "disintermediated" mercilessly. Such evolution made banks feel very nervous.

Regulating "Dividend"

"Banks' exploration in the internet finance is regulated strictly by the regulatory department, therefore, our boldness in innovation or enjoyment of regulatory

---

[6]Yu [3].
[7]Yu [3].
[8]Ma [4].

Table 2 Internet finance's main features and its corresponding sub-divided types of business

Business features	Subdivided types of business		Enterprises/product example
Through new technologies, tap the customer information and manage the credit risk	Price parity of financial products		Rong360, haodai, 51credit, Bankrate
	E-commerce platform Loan principal credit assessment		Ali small loans jingbaobei, Jingdong IOUs
	The third party credit assessment		Allwin (online loan credit) Treasure Island, 3golden (supply chain financing credit)
On the basis of the direct point-to-point transaction, the finance resource allocation is carried out.	P2P loans		wzdai, s-rong, eloancn, Hongling Capital, lufax, ppdai, creditease
	Crowding funding		demohour, AngelCrunch, zhongchou, dream-more, star.taobao
The-third-party- payment-based money transfer	Mobile paid	Products	Alipay wallet, wechat payment, 1qianbao, online banking wallet, ICBC e-payment
		Technology	sonic payment, code scanning payment, NFC payment, password payment
	The third arty account balance payment		Alipay, tenpay, Unionpay online
	Payment through online banking channel		wechat payment, Alipay wallet
	P2P funds custody		ChinaPNR, YeePay

Source http://paynews.net/portal.php?mod=view&aid=27129

"dividend" is much weaker than that the internet enterprises. Firstly, banks emphasize the safety." Zhang Lijun, Deputy General Manager of Electronic Bank Department in ICBC and the person in charge of e-commerce, mentioned firstly the regulatory issue when talking about the premise of the banks' exploration of the internet finance.

China's commercial banks were regulated by various indicators in the growth (see Fig. 1). Those regulatory indicators meant the development border and regulatory cost to the bank, but constitute the regulatory "dividend" to some degree for the internet finance. For example, the increase of the bank's capital adequacy ratio meant the loan freedom was limited and the profit space was compressed further. Therefore, the bank was less interested in the projects with high risk and more earnings and leaves the opportunities to the informal credit market. According to the public information, the registered capital of Wenzhou Loan was 5 million yuan

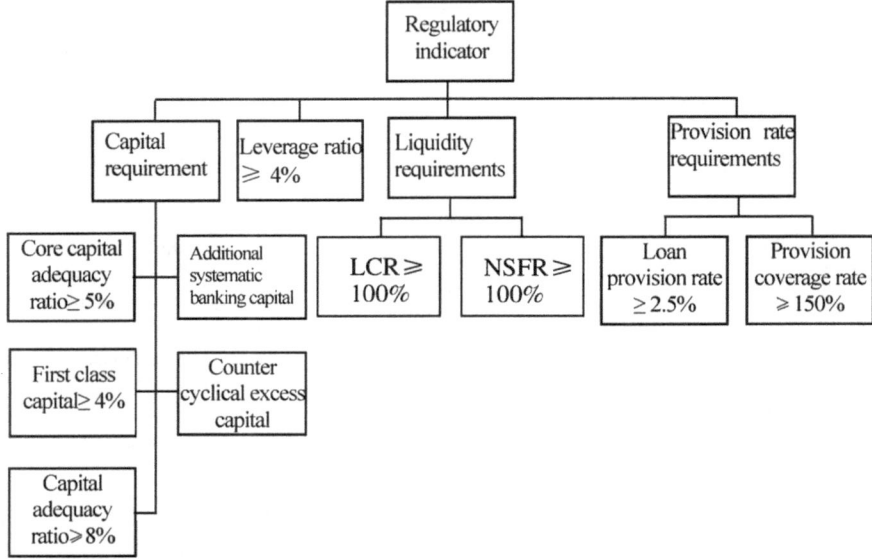

Fig. 1 China's current regulatory indicator system. *Source* http://www.cf40.org.cn/plus/view. php?aid=4119

and P2P was 1 million yuan, but their turnover could be increased to be hundreds of millions yuan, even several billion yuan in one year.[9]

One more example, if the term deposits in the bank were withdrawn ahead of the term, only the demand interest was obtained, however, the agreed deposits invested by the fund company in the bank enjoyed the "dividend" with "no punitive interest for the withdrawn deposits ahead of the term", namely the fund company, in case of the emergent redemption, could withdraw the undue agreed deposits, and the bank paid the interest according to the original agreed earnings and the relevant loss arising from the interest rate is borne by the bank. The agreement used to be approved by CSRC and CBRC. Because of such "dividend", enterprises like Alipay made commitment of T + 0 redemption so that a large number of customers with the liquidity demand for funds were attracted.

In addition, some third party payment, payment accounts and deposit accounts were confused so that the financial institutions border was fuzzy. The current regulatory situation showed that the debt and deposit businesses should be the concession of commercial banks or other financial institutions. But in the third party payment, those two accounts were confused, so the legal attribute of the accounts was not clear, no regulations on the account's interest was made. The above-mentioned two types of accounts could both take deposits and handle settlement without paying the reserve and accepting the liquidity management.

---

[9]"The Internet Finance Regulatory Gaps Breed Tempting Dividend, China Finance Information", January 8, 2014, http://www.cfi.net.cn/p20140108000586.html.

Nevertheless such "dividend" had aroused various parties' concern. In March 5, 2014, the internet finance was written into the current government report for the first time. The government would utilize more open and inclusive thinking and philosophy to guide the internet finance's standardized development when it controlled the internet financial risks. So the internet finance management had been included in the work agenda of parties concerned. For example, in March, 2014, People's Bank of China voiced that "if the offline financial business is put online and it should comply with existing laws and regulations and the capital constrains. Some unreasonable contract terms and conditions should not be permitted to be effective. For example, when deposits are withdrawn ahead of schedule or the service is terminated ahead of schedule but the interest bearing is still effective according to the original contract or the fee is collected according to the original standard."[10]

However, Zhao Xiaopu, researcher of policy study bureau in China Securities Regulatory Commission said, "now in the internet financial chain, except for some links that are regulated (such as the third party payment), other businesses are in the "three-non" state of non-threshold, non-standard and non-regulation."[11]

Challenges ICBC Faced

Now 250 third party payment enterprises had obtained the payment license in China, among which 100 enterprises were qualified to carry out the online payment business. Several top payment institution topped globally in the business transactions, amount and the number of customers. The businesses in which they were engaged included taking deposits, handling settlement, bank card acceptance, investment management, micro-finance and others, covering traditional bank business such as deposits and loans. They also stretched their tentacles to the financial derivatives such as fund and insurance. According to what Mr. Jiang Jianqing said *"the covering scope is one of the broadest all over the world and imposes greater pressure on commercial banks."*

In order to illustrate the pressure, Mr. Jiang Jianqing presented a group of data. In 2013, the commercial banks paid 1075 trillion yuan with 25.7 billion transactions. The third party payment amount was 9.2 trillion yuan with only 16.7 billion transactions. Since commercial banks couldn't conduct the cross-bank settlement, they were not linked directly and the settlement could only be made between commercial banks through the UnionPay. Therefore, in the cross-bank payment, commercial banks' transactions were just 2.1 billion with 15.3 billion third party payment's transactions.

---

[10]"People's Bank of China Re-launch an Attack at the Internet Finance and the Dividend of Monetary Fund Ends", Beijing News, March 25, 2014, http://www.nbd.com.cn/articles/2014-03-25/820059.html.

[11]"The internet Finance May Usher the 'First Regulatory Year'", China Economic Herald, January 23, 2014, http://www.ceh.com.cn/cjpd/2014/01/308527_2.shtml.

The reason for the internet company's rapid development in the financial field in recent years was that the financial services offered by them had characteristics of respecting the customer's experience, stressing the crossing marketing, advocating opening the platform and others on one hand. And on the other hand, *"the consumers in the internet financial ecosystem seize the information initiative greatly, and they actively look for products and services that they need so that they are not so loyal to their banks of deposit and unnecessarily follow the guidance of the bank salesperson, so the traditional relationship between banks and customers are spoiled. Meanwhile, banks still rely on the path which made them develop successfully in the past"* Therefore, to sum up those changes, Mr. Jiang Jianqing believed that *"Even if ICBC has made great achievement in the past informatization, but it is still challenged greatly."*

"The informatization in the past was based on the traditional way of thinking toward the internal and the past so that the bank's business was done through the computer way. The bank's nature has not changed, the bank's products have not changed, even the bank's process has not changed, the management mode has changed rarely and just the digitization manner was taken," Lv Zhongtao, general manager of ICBC technical department added.

"The present internet finance market is the consumer market. The survival of the fittest is that the one who satisfies the consumer demand can survive. Therefore, we are faced with the reform challenge that the decision-making way of thinking changes from internally. to Externally." Mr. Jiang Jianqing believed that this challenge was the internet thinking challenge ultimately.

So, on the basis of informatization, being faced with the internet finance challenge, how would Mr. Jiang Jianqing take the next step?

Actively Integrating into the Internet Finance

The internet finance is not simply defined as the financial business through the internet, and its essence lies in respecting customer's experience while offering the financial services, stressing the interactive marketing and advocating opening platform and rapid interactionetc. Therefore, the banking information technology cannot satisfy those requirements. From 2013, ICBC initiated the strategic transformation and it transferred into the informationization bank from the banking information technology. The transformation helps ICBC make the business decision and strategy with dependence on the data conversion instead of experience through information concentration, integration, sharing and tapping. Thence, we should cultivate the data analysis habit, attaching importance to the big data development and application, adhering to taking customer and market as the core, taking information as the guide, capital flow as the main line, logistics as the foundation, mobile bank with networking, convenience and customer self-definition as the behavior direction to reconstruct the banking system and turn the bank from the payment and financing intermediary to the comprehensive information intermediary.

—Jiang Jianqing

The banking information technology and the informationization bank were just reversed in their subjects and predicated literally, but what difference did they have in the end? Mr. Lv Zhongtao, General Manager of Science and Technology Department in ICBC presented "the banking information technology is the bank's traditional way of thinking on the basis of looking outside from the inside and the bank's nature is not changed. The products are not changed and the flows and the management model are rarely charged, too. However, the informationization bank is reversed, in which the decision is made from the perspectives of customers and markets. It is from the information perspective that the bank's business and management models are reconstructed, so its capability that information creates values is improved."

So, under the informationization bank strategy, how did ICBC improve its capability of information creating values? "Firstly, we should come to the internet finance," said Lv Zhongtao. The informationization bank strategy involved the implementation of four levels of infrastructure construction, operation, management and measures. Focusing on those four levels, ICBC designed 25 tasks. Some tasks were directly related to the internet finance. For example, at the level of infrastructure construction, the cloud storage and cloud computing technology were introduced to build up the foundation for big data. At the operation level, the platform related to the internet finance should be built up.

Beginning from the internet finance, how could they obtain the decision way of thinking from the perspective of customers?

Big Data

How to apply the big data? Mr. Jiang Jianqing had his own experience. From his point of view, the human brain was an application machine of big data. He often used his own machine to identify abnormal information in ICBC. In the past employees in Accounting Department in ICBC took the statement with more than ten pages to him for review. Although the statement was full of dense numbers in line and row, Jiang Jianqing could find out errors rapidly if there were any. One day, the person in Accounting Department who had sent the statement just walked away from Jiang Jianqing for a short while, Mr. Jiang found out a figure error, but the General Manager of Accounting Department said immediately that it was impossible since they had checked the statement for many times. With Mr. Jiang Jianqing's insistence, the Accounting Department checked it again. However when the figure that was corrected in a hurry was presented to Mr. Jiang Jianqing, Mr. Jiang pointed another error. Finally, the Accounting Department conducted a careful calculation calmly and corrected the error.

"My identification of errors results from the big data application in my brain," said Mr. Jiang Jianqing. "All data have their laws. As the data are accumulated to certain degree, their changing state becomes a psychological figure for you. Once you notice that the data fluctuation goes beyond its scope and you come to realize that the abnormality may occur instinctively." Mr. Jiang Jianqing said that his extremely high sensitivity of data resulted from the full data accumulated for a long time in his brain. "If we want to play with the big data, we should make analysis with full data."

After 30 years of development, ICBC had accumulated dozens of PB[12] of data. These data were stored in two types of database. The database was to store structured information. The information base was to store non-structured information. As to the non-structured data, ICBC stored them in addition to the traditional file form and taps and analyzed them on the basis of Hadoop.

ICBC's analysis and application of those data was deepening constantly with the appearance and gradual maturity of the big data technology. In almost all businesses and management modular in ICBC, such application was infiltrated. For example, ICBC had conducted the credit rating for 18 million customers in ICBC. At the instant when the customer swiped his card in the transaction, he would receive a short message. "You've just spent xx yuan and do you want to make installment?" If yes, as long as the customer answered the message, the just amount of consumption returned to his car, so the loan was completed.

This application had been built into a product—Ease Loan. "The difficulty of this application lies in how to select customers. Without the big data support, we couldn't dare to introduce the new product." Zhang Ying, deputy general manager of science and technology department in ICBC said. "We designed a model selecting customers. The system, in accordance with the model, adds or deletes customers dynamically."

In January, 2014, ICBC offered the application of credit card anti-fraud by making use of big data online. Abnormal transactions out of 4 million daily transactions with credit card were intercepted. In May, ICBC succeeded in intercepting a foreign transaction with the stolen credit card. That day the card of a customer with high net value in ICBC was swiped continuously in some Arabic country. As soon as the system caught such information, the account was frozen decisively. "We conduct interceptions through two ways. Firstly, it is to apply rules and variables under rules. More than 100 variables are tapped through data and they can identify risks truly. Furthermore, the variable tapping is a dynamic process. At intervals of time, data are rescreened according to new fraud means. Secondly, the risk scoring is carried out to transactions. Spatial and temporal analysis r are involved here." A responsible person in ICBC Shanghai R&D Center said that "We require the system to operate all those rules within 50 ms, because we have to make sure that swiping customers have good application experience, which of course raises more strict requirements to the system in terms of the real time processing capability and capacity.

---

[12]PB is a unit of data storage capacity, which is equal to 2^{50} bytes or about 1000 TB in numerical value. TB is a unit of storage capacity in computer, and it is equal to 2^{40} or nearly 1 trillion bytes. Raymond Kurzweil, futurist, extends the PB's definition: the human being's functional memory capacity is estimated to be 1.25 TB, which means that 800 human being's memories is only equal to 1 PB.

As to ICBC's big data construction and application, Mr. Qian Bin, General Manager of ICBC Data Center (Shanghai) believed that there were still three shortcomings: "first, the application of the structured data is inadequate. We have more than 400 corporate customers and 400 million personal customers. Everyday there are data of more than 200 million transactions, which are not sufficient to support the business operation; second, our capability to collect and analyze the non-structured data need to be improved; third, in some extent, our information is fragmentized and it needs to be integrated and promoted." Mr. Qian Bin also pointed out that ICBC's measures of integrating the front platform and sharing the middle and back platforms were to break boundaries of professional departments and achieve the systematic integration. "This is an important foundation to realize big data processing, otherwise there will be fragments and separations to the information."

Setting up the Internet Financial Platform

Payment

Through the payment platform, ICBC hoped to offer all capital settlement businesses supporting B2B and B2C. ICBC had the historical accumulation for the payment platform.

As early as in February, 2008, ICBC opened the payment function in the online banking. In November, 2011, ICBC launched ICBC e-payment.[13] On such basis, in 2014, ICBC began to build up the internet payment platform. Hou Benqi, General Manager of ICBC Electronic Bank, said *"this payment platform's characteristics are: first, unified. All branches and different payment businesses use different platforms before, but now a unified platform is built up; second, in the terms of application, the customer's experience should be ensured; third, the mobility should be supported."*

ICBC e-payment was one of important products in such platform. In March, 2013, the new version of the product was launched so that customer entered the SMS verification code received by the reserved mobile number and completed the online shopping, transfer and bill payment within 3000 yuan for each transaction on the mobile payment. But in 2011, if the user used this product to conduct payment, he had to carry out his verification by the way of "mobile number + the last six figures in the bank account + the dynamic password".

"Don't look down on the change of the verification, which was argued repeatedly inside ICBC and it was the tradeoff result," said Zhang Ying. *"In fact this is a reflection of our transfer to the internet way of thinking."* According to

---

[13]ICBC e-payment was a payment in the PC terminal or mobile terminal to be promoted to satisfy the customer's small amount payment demand.

Zhang Ying's introduction, the former verification way in ICBC was out of consideration of absolute security. Therefore, when the password verification way of thinking was proposed inside ICBC, business departments were against it immediately. *"How about money being lost? Who is responsible for such loss?"* However, the user's experience in the former payment verification way was obviously poor. It was troublesome. As the third party payment product had helped the users experience the simple and fast way. If ICBC continued to keep the former payment way, how did it attract customers? *"Pressure in the market forces ICBC to work out the compromised manner in security and convenience and break the traditional thinking inertia,"* said Zhang Ying.

In addition to the e-payment, ICBC also offered the non-ICBC customers the payment services. In April, 2014, ICBC launched the cross-bank payment service. The user can also use other banks' card to carry out payment in ICBC ATM or other terminals. Furthermore, in July the payment platform added the payment service without card.

Comprehensive Charge Payment

In the comprehensive charge payment platform, ICBC played an intermediary role in linking merchants to customers. ICBC would turn to the "payment + credit + information" intermediary from the "payment + credit" intermediary, and this charge payment platform was an indispensable part, which could bring about the transaction information about the people's livelihood for ICBC. In fact, the third party payment had also been seizing the middle position. Therefore ICBC felt greater pressure.

All branches or online banking of ICBC had already offered the charge payment on the people's livelihood such as charges for water, electricity, gas, tuition and medical expenses. But the former charge payment systems scattered in different branches of ICBC and it was not favorable for the information integration and overall planning and management. So ICBC hoped to build up a comprehensive charge payment platform. In this platform reconstruction, the user's satisfactory charge payment experience was what ICBC had been trying to achieve. "In this charge payment market, we have advantages originally and moreover we are able to develop the market rapidly and to provide more convenient service for customers. For example, the customer can pay charges without the online banking and he may pay charges with other bank cards," said Zhang Ying.

Online and Offline Integration POS Business Management

"This is what ICBC is doing and it is predicated to be launched within the year." Zhang Ying said, *"The purpose of building up the platform is to attract the offline*

customers and stabilize the traditional offline advantage of ICBC. Meanwhile, ICBC would like to share the customer's consumption information and transaction information with businessmen."

The offline resources were mouthwatering for the internet enterprises to enter the financial market and they were eroding the offline market from the online market.

Zhang Ying said that the platform purpose could be achieved by embedding the location service function in the ICBC mobile banking so that the customers found out the consumption places nearby through the location service, then ICBC credit card or two-dimensional code were used to carry out the online payment or the offline payment by swiping card, and they could spend money offline. *"What we are doing is to provide the guiding service for the businessman and lead the customers to them. Even ICBC can also make different policies to attract different customers for various businessmen. For example, in a shop, the customer can enjoy the installments with low interest or even without interest, so the customer may visit the businessman more frequently,"* said Zhang Ying. Various banks were putting their POS machines there, how to make the businessman to choose ICBC? *"Beside those online guiding methods, we can also offer discounts for the businessman in the terms of corporate business, and the convenient financial services attract them"*, answered Zhang Ying.

Instant Message

The instant message platform in ICBC was a social product like Wechat. With it ICBC would solve two problems: first, the downward communication with customers in ICBC. In the past the active marketing to customers in ICBC was carried out through SMS and ICBC had to pay hundreds of millions of RMB for SMS annually. With this platform, ICBC can reduce partial costs. Second, this was more essential. The personalized and fast service should be provided for customers. at that time, the services offered for customers by ICBC were based on standardization and they were passive, i.e. the customer called or used other channels to present their complaints and ICBC solved them. On the platform, each customer manager could organize his customers to set up a group so that they had chances to communicate with each other within the small size and their communications were not limited to time. The customer manager could also provide the target services around the clock. In addition, in this platform, the customer could establish his own interactive sub-group.

Why should ICBC set up an instant message platform like Wechat independently but not make use of Wechat? *"In this platform, there are many financial activities that are not offered by Wechat, because it involves the security and privacy issues. This platform is also different from Wechat essentially: the Wechat focuses on chatting with the additional payment function; but our platform focuses on the financial service and one of menus is chatting,"* said Zhang Ying.

In April, 2014, ICBC launched the 1.0 version instant message, then all messages about ICBC received by customers were sent by this platform. As of the mid-July, about 3 million verified customers had gathered on this platform. Registration was necessary for Wechat, but only one ICBC card was needed to open the instant message. With an ICBC card, the customer was considered to be able to enjoy such instant message service automatically, but the customer needed to confirm unilaterally whether he was willing to use the function or not.

E-Purchase

Why Do It?

The e-purchase platform, in accordance with what Mr. Jiang Jianqing said, was a "cross-sector" action for ICBC. "ICBC can do nothing but grasp customers' commercial information, as some of the third party payment companies don't share such information with us, which brings about risks to us."

In order to avoid risks, ICBC decided to involve itself in such unfamiliar sector. On January, 12, 2014, the e-purchase B2C platform was on line.

"Our aim to build up this platform is completely different from other e-commerce enterprises. We don't have the commercial interest demands, and we just want to use this platform to link our corporate customers and personal customers to establish a commercial ecosystem. What we are concerned about is the payment, financing and other information brought about by mutual transactions in the platform." Zhang Lijun, Deputy General Manager of Electronic Bank Department in ICBC and the person in charge of the e-purchase business said, *"You have to pay if you conduct transaction in the platform. After the payment is conducted, both sellers and buyers have possibility of financing. In the back end, we make use of big data analysis to evaluate the customer's credit in advance. Once the customer has the financing demand, the financing can be ready immediately. In addition, through this commerce platform, we are able to capture the customer's habit information. Therefore, the platform becomes an information node for us to collect big data. In one word, ICBC uses the internet thinking service to serve the whole ecosystem."*

Current Status

As of June, 25, the e-purchase customers reached 4.36 million with the total transaction volume nearly 6 billion yuan. The aim of the e-purchase was to have the transaction volume reach 15 billion yuan in the end of the year. But Mr. Jiang Jianqing thought that *"It is too little, and ICBC should achieve 100 billion yuan as early as possible."* From his point of view, this was not a demanding requirement. *"The average transaction volume in the single transaction is about 2000 yuan, and it is about ten times Taobao. If we are able to attract 10% of our customers to carry out shopping, this is not a very small scale."*

The average single transaction volume in the e-purchase platform reached 10 times Taobao and this is because ICBC made the "three-famous" positioning strategy for it, which was the famous e-commerce enterprise, famous commodity and famous store. ICBC required those enterprises that intend to land in the e-purchase platform should at the top four or five in its sector and ICBC also proposed higher requirements in the brand's popularity, e-commerce enterprise's experience and service capability.

Could ICBC attract e-commerce enterprises with so much higher requirement? Zhang Lijun disclosed that the day when the financial e-buy was on line, more than 200 e-commerce enterprise landed in the platform. In the end of the year, 2000 enterprises would station in it. "Now more and more enterprises want to land in the platform, but we must screen them strictly according to our requirement." Zhang Lijun said, some well-known breweries mentioned that "now we are not short of loans, what we are short of is who can help us sell our products", so did many other corporate customers of ICBC "This demands us to change to offer the comprehensive services such as "marketing + finance" from the past financial services."

The services concerning marketing offered by ICBC was very attractive to some customers. For example, one well-known fertilizer company in Guizhou heard that ICBC launched the e-commerce service on line and it decided to open a B2C direct sales store in the platform and tried the online direct sale model although it had already taken channels for sale.

"In the offline channels for sale, there are puzzles of many marketing nods, long process and unsecured quality and price. We want to do e-commerce, but if the ICBC e-purchase is not put online, we would not take actions so fast. Actually we are not familiar with other e-commerce platforms and we don't have clear ideas how to deal with them. But we know ICBC well." One executive in the company explained his reason of taking the platform as the priority.

The company was on line in the platform in May. The one-month sale volume was more than 10 million yuan. Although this amount of sales was too little for the company with the annual sales of several billion yuan, the company was running the direct sale store with effort. Their next step was to station in the B2B business platform and hoped to combine the business model of the B2C and B2B.

The reason we decided to invest in the platform was we want to cure for the chronic disease in the traditional channel and explore a new sales model; Meanwhile, we want to make use of ICBC's outlet resources in rural areas to expand the market, said the executive.

The e-purchase not only attracted corporate customers but also non-corporate customers. LeTV was one of the latter. It stationed in the platform while it was put online. "We station in the platform, because we eye on its many quality customer resources. Meanwhile it has very precise information so that we can conduct the targeted marketing." Zhang Jing, the person in charge in the LeTV market said. LeTV also stationed in Jingdong and other e-commerce websites. From Zhang

Jing's point of view, at the present stage, the e-purchase couldn't compete against those websites in traffic, but "we eye more on its special means of tapping users. In addition, its platform positioning can help us to build up the brand."

Different Methods with Other E-Commerce Enterprises

Being different from the non-banking e-commerce, in the birth process of e-purchase, both the e-commerce law and the bank's advantages were stressed.

Other e-commerce enterprises depended mostly on the price for profits, but in the "three famous" positioning, the "quality + reputation" was emphasized and the organic combination of the commodity quality, the seller's reputation and ICBC's reputation should be realized. Such positioning determined that the price war should not be launched. So how to attract consumers? Zhang Lijun answered, "we made an investigation that consumers of online shopping have changed their focus on cheap price to gradual attention to the commodity quality and seller's reputation. Existing e-commerce platforms out of historical reasons are hard to have a leap-forward in the brand. This leaves spaces for enterprises like ICBC who have the well-known brand and no e-commerce history before."

Another different practice for the e-purchase was that it wanted to realize five combinations in the first year: combination of B2C and B2B, direct marketing and sales agent, finance and non-finance, PC terminal and mobile terminal and domestic business and overseas business.

The first combination would start in the second half of 2014. At that time, with the ICBC B2B platform being online in e-purchase website, according to what Zhang Lijun said, the e-purchase would turn into the "B2B2C platform". What was the B2B2C? He gave an example. A vehicle manufacturer sold the vehicle to the dealer in the platform (B2B), and the dealer sold the same vehicle to the individual in the same platform (B2B2C); or some brand owners opened the direct sales store in the platform. Meanwhile he worked together with the dealer (B2B) to sell commodities to the individual (B2B2C). "The combination's purpose is to realize our supply chain's financing", said Zhang Lijun. "The back stage's data of the bank are interlinked. We can conduct the trade financing for the dealer according to the supply chain transactions, and also for the final consumer. When the whole supply chain is promoted to be running organically, enterprises, personal customers and our bank all gain benefits from it."

The B2C-B2B combination would result directly in the necessity of the second combination. Enterprises could conduct direct sales and they could also introduce their own dealers or agents.

The combination of finance and non-finance was the position of the e-purchase commodities. Finance meant the bank's private investment, fund, insurance, precious metals and others. Non-finance covered commodities and services. The services referred to charges payment for livelihood. Zhang Lijun believed that such business management not only played the role of the bank's advantage but also derived the sale of non-financial commodities through financial commodities. Meanwhile logistics could also be avoided, and "logistics is not our strength."

The combination of PC terminal and mobile terminal was of great significance. All e-commerce enterprises had been exerting greater efforts in the mobile terminal and there was a tendency that any e-commerce enterprise that had the mobile terminal could enjoy the global markets. However, no enterprise had won till present. Hence, the mobile terminal bred opportunity for banks who came from behind Even the e-purchase platform was launched for just half a year, the mobile terminal of the financial e-buy APP had been launched and the iPad terminal of the financial e-buy APP would be launched as well. Zhang Lijun said, "next, on one side, we would combine the PC terminal and mobile terminal and optimize it continuously; on the other side, APP and ICBC mobile banking should be combined organically to promote the e-commerce development in the mobile terminal."

The e-purchase platform also planned to make use of ICBC's global resources to develop the overseas e-business market in the second half of the year, namely the domestic customer could purchase foreign commodities in the platform and the domestic commodities could also be sold to foreign markets.

Infiltration of the Internet Thinking

The cross-sector statement of the e-purchase by Mr. Jiang Jianqing referred to not only cross to the non-financial sector but also "reflect characteristics of the e-commerce". The biggest characteristic of the e-commerce was the customer's experience first and it was greatly different from security and safety that the bank was pursuing of.

Zhang Lijun was a person with the "cross-sector" ability. When he worked in ICBC, he did the human resources job. Later he served as the league secretary in ICBC and then he assumed the responsibility for the customer series in the e-banking department. When he was assigned to do e-business, he said wryly "except for understanding the human resources, I know nothing about the e-commerce." Nevertheless, he said that he was doing the e-commerce program with a mentality as a "waiter" and serve for customers actively.

Zhang Jing in LeTV had greater experience for this. The people of ICBC e-purchase platform contacted him actively for several times and intended to help LeTV have an exclusive insourcing show, namely an e-commerce marketing program just for the large ICBC corporate customers. In this program, several hundreds of LeTV TV sets could be sold out one day. During the cooperation with ICBC, Zhang Jing felt surprised that "ICBC's idea and exploration effort for new model are greater than other banks and even stronger than e-commerce companies." Recently, Zhang Jiang was discussing a new cooperation model with ICBC, "Originally I never thought of adopting the online and offline integration model, and ICBC inspired me and offered me very good resources support." Because of ICBC's active innovation, Zhang Jing planned to carry out deeper cooperation with the financial e-buy platform.

The e-purchase platform couldn't depend on its own effort to practice its philosophy of customer first, and it should need other departments' cooperation. For example, its customers Haier, TCL, etc., had their mature e-commerce order system and transaction system. When those customers stationed in the platform, the platform needed to adapt itself actively to their systems. However, in the system development, ICBC adopted generally the open standardized interface and anchoring manner. It was impossible obviously for those mature enterprises to anchor to ICBC. Therefore, it brought about a personalization challenge for them. Of course, with the challenge, the internet thinking infiltrating from externally to internally was also approaching.

Action Breaking Through Concepts

"In the e-purchase platform, we used a large number of young people and encouraged them to carry out iterative innovation and development and set up platforms for them. We also encouraged them to conduct more attempts and make correction if any problem occurred. It was rare in our past practice. In the past we got used to pursuit of being reliable and safe. As one program was implemented safely and soundly, we launched it to public," says Mr. Jiang Jianqing.

Besides, in the introduction of sellers, the e-purchase platform also broke the traditional concepts in ICBC. ICBC implemented the strategy of "introduction first and evaluation second". If sellers introduced did well, they could continue. Otherwise they should be off the shelf. On the traditional banking business, Zhang LiJun said, *"All diligence investigations should be conducted firstly. Only those enterprises being evaluated as Grade III A can be introduced."*

"The f e-purchase is one of our counter-measures against the internet's challenges. Through this platform, we are integrating the capital flow, logistics and information flow. This is an attempt that it is completely different from the past commercial banking model. The process offers us much inspiration, so we have been still in exploration in this aspect." As he talked about it, he already stepped into the internet finance pool. Ahead he saw a group of aggressive rivals who were running toward him, how would he deal with them?

Digitization Reshaping Competitiveness

Funds set off from the banking system originally and return to the bank after various flows. Therefore the bank should control the capital flow, logistics and information flow essentially. If we are able to open up three flows and play the role of the information intermediary in the process, we can transform information and create values. I even predict that one day the value that we gain from the information transformation will be greater than that we obtain from the capital transformation. Then we are the true informationization bank.

—Jiang Jianqing

The information intermediary, from Mr. Jiang Jianqing's point of view, was the most capable role that ICBC could play in the internet financial ecology and it would also become the differentiation competiveness for ICBC to compete against rivals. So how could ICBC stand in the position of the information intermediary?

On one hand, ICBC should integrate itself into the internet finance actively to open up the three flows. On the other hand, in accordance with what Mr. Jiang Jianqing said, *"The big data technology is applied to create information competitiveness."* To this end, in 2013, ICBC initiated the "large company" strategy. In 2014, it initiated the "large retail" strategy, "large asset management" strategy and the "big data and informatization" strategy. Those strategies were molecular strategies of the informatization bank strategy in recent two years. In the process of being infiltrated gradually by the internet way of thinking, ICBC re-built the business mode and management mode from the information perspective through implementing those molecular strategies.

Initiating the "large retail" and the "large company" strategies was to collect, integrate and process information with a unified customer view and repay customers in a way of unique competitiveness with values created by information on facing the personal and corporate customer terminals. The "large asset management" strategy supporting those strategies provided shells for the "large retail" and "large company" strategies—satisfying customers' demand for the financial wealth management products in the internet ecosystem.

The implementation of the "large retail" and "large company" strategies had improved the integration of big data and enriched the sources of big data. Conversely, tapping and analyzing big data not only enhanced the "large asset management's" capability to innovate products but also improved the further implementation of the "large retail" and "large company" strategies. Therefore, the "big data and informatization" strategy should step side by side with other strategies to form the mutual and sound promotion.

Value created by information should become competitiveness in the end, which didn't work without the credit risk control, so firstly ICBC should make use of the big data technology to improve the credit risk control mode so as to make it play a role of powerful protection continuously in the internet financial ecosystem.

Credit Risk Control

In the internet financial ecosystem, most customers that ICBC was facing were small and medium enterprises, even small and micro enterprises. Those enterprises were different from ICBC's former customers. In accordance with what Mr. Nie Dazhi, Deputy General Manager of Credit Management Department in ICBC said, "the concealment of risk in those enterprises is greater, and the associated risk arising from associated transactions is hard to be identified and prevented".

Facing those enterprises that were many in quantity but not large in size, the internet (enterprise) finance pioneered a mode to control risk by the big data, which was from e-commerce transactions, social networks, etc.

Therefore, Mr. Jiang Jianqing required relevant departments in ICBC, "the experience of the internet finance should be combined with the bank's current risk control system and the big data should be used to strengthen our existing risk control management system."

The embryonic form of the ICBC credit risk control system was the legal person credit management system that was developed and went into operation in 2003 and the individual credit management system that went into operation in 2006. In two systems the corporate or individual customers were taken as the center, the business process as the main line and the risk control as the focal point respectively to realize the credit business real-time operation, the data centralization management and the unified risk control.

"Whether in the past or now or in the future, the risk control in ICBC is based on the data management, and they are different in degree. A company comes to the bank to apply for credit and it has to go through two steps of rating and credit granting. Credit is granted in accordance with its rating. What is based on to rate the company? It was based on data before, which came from the financial statements or the investigation on site, etc. But now more and more data can be collected and in the future even more." Mr. Zhang Ying, Deputy General Manager of Science and Technology Department in ICBC said. In order to have more data for rating, ICBC had been breaking the barriers between departments by the "large retail", "large company" and "large asset management" strategies, integrating the data scattering in various business departments, and expanding the channels for collecting external data. Meanwhile, ICBC focused on collecting information about the enterprise business management and settlement. The third part data such as data of People's Bank of China, industrial and commercial bureaus and Ministry of Public Security was also collected by them.

In this risk control system, ICBC set up the Credit Regulatory Center in the head office in 2014. *"The Credit Regulatory Center conducts a centralized and dynamic monitoring of the ICBC's customer credit risk. It aims to carry out the identification beforehand and present the risk warning. It conducts the beforehand monitoring through data tapping and model budget processing in a remote distance."* Mr. Nie Dazhi said.

The system relied on the multidimensional customer's information collected through various channels, and the relevant model identification analysis was established to control relevant risk points such as the abnormality of the capital chain, even the abnormality of the legal person's behaviors and others, which warned relevant business employees in ICBC. For example, it was known from the transaction payment system that a company had not paid its water and electricity bills for three months, the message would be transmitted to the customer manager as a beforehand warning. Such beforehand warning was presented before, during and after the loan was made, and the site monitoring was unnecessary.

The big data-based credit risk control system was not only applied in the credit business but also in the investment banking and other business. In 2014, when the "large asset management" strategy was finalized, ICBC conducted a docking of the relevant business management system and the credit management system and achieved the unified risk control of the credit and acting investment business. Meanwhile, in order to promote implementation of the "large company" and "large retail" strategies, ICBC integrated the legal person credit management system with the personal credit management system to build up a global credit and acting investment management system covering the domestic and foreign legal person credit and personal credit so that the integrated credit risk management was achieved from the personal credit to the legal person credit, from the business in the statement to the business beyond the statement, from the domestic institutions to the overseas institutions and from the credit business to the acting investment business.

As of June, 2014, the personal credit in the system had been in full operation in ICBC, the overseas institutions' various credit business was under operation and management in the system, too. The domestic institutions' legal person's credit and credit-typed acting business would be put into pilot operation in the second half of the year.

Large Asset Management

The "large asset management" strategy in ICBC mainly involved asset management, acting transaction, market analysis, private investment consultancy, asset trustee and other business. Those businesses were the major part of the intermediate business income sources in ICBC. The pressure arising from the interest rate marketization and the opening up of the financial market forced all commercial banks to take the intermediate business as one of the key fields for survival in the future. Therefore, in the ICBC's informatization banking strategy, the "large asset management" became one of essential tasks in re-integration.

"The asset management business mode is different from the traditional banking business mode essentially," said Chen Xiaoyan, Director of Asset Business in ICBC. *"In the asset management business, the customer entrusts his interest and right to the bank, so our professionals must create values for the customer with diligence so that the commission concerned can be charged, otherwise the business model of the asset management could not be built up."*

In 2009, ICBC established Asset Management Department as the first in all commercial banks. In 2014, with more competitors in the asset management market, ICBC management made a "large asset management" strategy resolutely, aiming to "grasp direction of the social financial asset migration, the customers' demand changes, regulation reform requirement and financial innovation in the large asset management times. In accordance with the essential attribute and development law of the asset management that' Being Entrusted and Acting Private Investment', ICBC sets up a 'large asset management' business model covering all customers, investing all markets, innovating all products and developing whole value chains." Chen Xiaoyan presented a further explanation.

The key for ICBC to implement the "large asset management" strategy lied in its cross-integration and mutual promotion with the "large retail", "large company" and "big data and informatization" strategies. On one hand, the "large retail" and "large company" strategies provided the product markets and channels for the "large asset management" strategy. On the other hand, the "large asset management" strategy, through offering products and services, improved the marketing effect of the first two. Finally, the "big data and informatization" strategies could support the "large asset management" business to achieve the rapid innovation of products, the refinement and development of the product service channels and the targeted improvement of the customer service experience and the improvement of the risk management and control level.

For example, the R&D Department in ICBC sorted out and analyzed relevant business products. According to the design property of modularization and formation, it built the property design of various types of business products into the "parts" of the system. The business people could allocate a variety of new products flexibly and rapidly on the basis of those "parts" according to customer s' demand characteristics. The system had been applied in the "large asset management" field and relevant people innovated the private investment products rapidly to satisfy the customer s' demand.

Relying on the combination investment business mode, Li Xiaofeng, director of ICBC Beijing Xiangpu Subbranch, successfully turned an enterprise engaged in the international business into his customer recently. The enterprise imported commodities from overseas and purchased foreign exchanges from foreign banks. Since it was much cheaper for this enterprise to buy foreign exchanges from foreign banks than from Chinese banks, Li Xiaofeng had had no hope to acquire this customer even if he wanted. However, later, he could help the enterprise gain more profits through a series of investment products combination.

If the enterprise intended to pay $1million for the overseas company, it had to purchase $ 1million with the equivalent value of RMB from foreign banks and then could make remittance. But with help of the investment business combination offered by ICBC, Li Xiaofeng persuaded the enterprise to make investment with RMB with which the enterprise intended to purchase foreign exchanges, for example, the enterprise might purchase the investment product with the three-month term, then it could use the investment product as collateral to apply ICBC for three-month USD loan and pay $1 million to the overseas creditor. Because there was an interest rate gap between RMB deposit and USD loan, after three months, the enterprise could use the equivalent RMB of $1 million and its investment profits to repay the $1 million loan and its interest during the same period. In the product combination, the interest rate gap between RMB and USD was used to conduct the cross-currency deposit. The combination not only helped the enterprise reduce the foreign exchange cost but also gained the profits of the interest rate transaction. More than that, Li Xiaofeng also made use of the instruments of forward foreign exchange,[14] currency swaps[15] and others to help the enterprise realize the risk aversion of the exchange rate fluctuations.

Li Xiaofeng's success of making best use of the combination was based completely on the scale advantage of the "large asset management" strategy in ICBC. "Only if the combination product is provided in accordance with the customer s' comprehensive demand can the investment be made in the domestic and foreign markets and bring about the scale advantage."

Large Retail

Like all commercial banks, the retail business in ICBC had provided financial services for personal consumers from ICBC's perspective. However, according to what Mr. Ren Ximing said, who was the Deputy General Manager of Personal Financial Business Department in ICBC (hereinafter referred to as "Personal Financial Department"), the "large retail' concept was completely different from the retail one. The service was offered from the customer's perspective and on the basis of his demand. "This is the result from the inspiration by the internet way of thinking."

Thus, Ren Ximing said that "in accordance with the 'large retail' strategy, ICBC broke through the financial service scope and combined the financial service with the non-financial service organically. Centering on the customer's basic necessities, travel, entertainment and shopping, on one hand, ICBC offered the financial service; on the other hand, it offered services of settlement and payment. For example, ICBC could provide the bill payment service and bank-hospital all-in-one card[3] service."

In addition, t Ren Ximing said, under the "large retail" strategy, ICBC was no longer limited to concern with its own customers, and it would show its cross-bank concern over and offered service for the non-ICBC customers.

In the "large retail" strategy, in 2014, ICBC set up the unified customer view, namely through integrating and upgrading the personal customer information

---

[14]Forward foreign exchange refers to the forward exchange contract signed with customers, in which the foreign exchange currency, amount, exchange rate and term are agreed in the future settlement of exchange and sale of exchange. When foreign exchange earnings occur at maturity, in accordance with the agreed currency, amount, exchange rate in the forward foreign exchange contract, the settlement of exchange or the sale of exchange is carried out. As a emerging financial instrument, the forward foreign exchange business offers functions of risk aversion and fixed exchange rate. Through this business, customers can realize their aims of risk aversion and hedges against inflation in the business activities involving the foreign exchange investment, financing, international settlement and others, etc.

[15]Currency swap means generally that the forward currency is purchased when its spot currency is sold, namely, currency A is sold at sight and currency B is purchased. Meanwhile currency B is sold and currency A is purchased. Currency swap is an effective means to fix the exchange rate risk and debt cost.

system and the marketing management system, all information about the customer in ICBC was collected under one customer's name. In such way, on one hand, the credit risk could be prevented effectively; on the other hand, ICBC could provide more precise services and marketing for customers. For example, when a customer swiped his card in the queuing machine in the ICBC branch, a queuing slip was presented, showing the customer type with marks and his product information. Then the business manager in the bank lobby and the teller could know the information at the first time and offered him targeted demand inquiry and product marketing.

Obviously, a series of other strategic support was required to implement the "cross-border" of the business and the customer in the "large retail" strategy. For example, if the customer demands for the customized investment product, the "large asset management" product support was needed, but to satisfy the financial and non-financial needs of various customers and improve their experiences, the internet financial platform support was needed, which was developed under the "big data and informatization" strategies. Furthermore, under the support of the credit risk control system on the basis of big data, the "large retail" system could provide customers with convenient loan products of the ease loan and unsecured small business credit.

The "large retail" strategy was promoted by ICBC Retail Financial Business Promotion Committee. The chairman of the Committee is Mr. Yi Huiman, President of ICBC. Its members were from more than 20 business departments, including Asset Management Department, Company Department, Precious Metal Department and others with Personal Financial Department as its leading department.

Under the framework of the "large retail" strategy, Personal Financial Department was the customer marketing department, and other departments including Departments of Bank Card, Personal Banking, Precious Metals and Asset Management focused on offering products.

As one of key measures of the strategy, ICBC carried out the pilot work of "the retail business pioneering development and reform" in more than 80 branches all over the country. Under the guidance of the large retail strategy, ICBC had been exploring the new retail business development model. To those pilot branches, the Retail Financial Business Promotion Committee granted certain innovation space and freedom. "The freedom used to make relevant departments in the headquarters of ICBC felt shocked and worried, but we convinced them with reform so that the pilot work can be carried out with bold innovation in a relatively relaxed environment," said one director of Personal Financial Department in ICBC.

Large Company

The initial "large company" concept in ICBC initially came from the customer's demand changes. Firstly, the cross-province and cross-border businesses were more and more popular for large companies. So the demands for the linkages between the parent company and its subsidiaries and between one subsidiary and another one grew up. ICBC used to just offer services for local subsidiaries of the group customer. Then such way of services was no longer suitable. Meanwhile, the corporate customers had been expanding its financing channels constantly, strengthening its financial and capital intensification management gradually and improving the employee management and services gradually, what the customer needs was not just the credit and it also needed financial services in the terms of investment banking, cash management and annuity management. In fact, some businesses of the "large asset management" also came from those demands. In addition, with the enterprise's attaching importance to the whole strength of the supply chain, the supply chain finance became the enterprise's important financial service demand progressively.

All those demand changes forced ICBC to apply the "large company" strategy to change the original public business model.

The so-called "large company' strategy, as what Xiong Yan, Deputy General Manager of ICBC Corporate Financial Business Department (hereinafter referred to as "the Corporate Department") explained, provided a package of comprehensive financial services in China and abroad and the domestic and foreign currencies for the customers of large, medium and small companies through all products, whole process and whole chain marketing and through the online and offline channels in accordance with various financial service demands of different types of customers." Therefore, she emphasized that "the core of the 'large company' strategy is linkage".

Just like the "large retail" strategy, it was the highest Financial Business Promotion Committee in the head office of ICBC that promoted the "large company" linkage. The Committee's chairman also was Yi Huiman. 29 departments out of 40 were members of the Committee in the head office and the Corporate Department was responsible for implementation.

The Corporate Department conducted the horizontal linkage of various product departments to satisfy diversified financial needs and the vertical linkage of branches in China and abroad to offer customers the cross-region financial services. The Company Department also established the linkage contact person network covering all branches, focusing on the cross-region linkage. Through Wechat and other channels, relevant linkage people were able to contact real time, update information, keep close contact and push forward the linkage service in a coordination manner.

In 2014, the linkage of the public and private business was a key task for the linkage of the "large company" and the "large retail" strategies. When the person in the company department visited the VIP customer, he had to bring his colleague in the personal capital department, and when the person in the personal capital department visited the personal customer and an executive of the company, he also promoted the corporate business to them In order to encourage such promotion, ICBC put relevant indicators of the corporate-private linkage into the assessment requirements. In Li Xiaofeng's branch, if the employee succeeded in marketing the non-duty business, he would be rewarded with bonus twice high as the standard one. "I am doing my job in such way to encourage the corporate-private linkage," said Li Xiaofeng.

At the end of 2013, like the "large retail", at the system level, ICBC re-built up the marketing system of the company and the legal person customer supporting the "large company" development, which also unified the customer views. For example, all information on the PetroChina scattering in various business systems and branches before should be integrated into the PetroChina view to form the full data about it and offer better services for it through the big data tapping.

Although all strategies in ICBC were advancing in such mutual promotion, the non-financial company's internet "spoiling" actions were also more and more fierce. Those spoilers were seizing markets in the mobile payment field. For example, in May, 2014, Alipay launched the "Future Hospital" program, transferring the registration, payment and prescription pricing to the Alipay platform, which used to be carried out in the hospital. In August, 2014, Tencent and Sinopec signed the cooperation framework agreement, in which both parties agreed to conduct cooperation in the fields of mobile payment, O2O business, map navigation, big data applied in the cross marketing.

However, Mr. Jiang Jianqing seemed not to be anxious about it, "sometimes I am thinking quietly that the ginkgo tree can live for 1000 years, and the turtle can live for 200 years because of slow speed. In the Periodic Table of Elements, the most inactive element can exist for longest time. The banking industry also has a history of 3000 years. It is good to grow up slowly. So even if we are changing ourselves, we should not be too excited, we should set aside time for observation, remove those unstable factors, take good factors into consideration and turn them into stable and normal factors."

In the internet ecosystem, Jiang Jianqing, with such a mentality, was facing the eyeing "spoilers" who were several 12-years younger than him. One party was calm and steady and the other was radical, who would live longer?

Case Analysis I

Addressing the Challenge from "Black Swan" Internet Finance with "One Body Two Wings"

Lao Guoling,

The case study *ICBC in the Digital Age* includes various theories and events related to the risks of Internet finance, where uncertainty has become the "norm". This case study considers approaches to "reconstructing" bank systems and "reshaping" the

Doctor, Associate Professor at the College of Business and Director of the E-commerce Centre of Shanghai University of Finance and Economics.

operation and management models of banks from the perspective of information transformation with a "one body and two wings" structure.

One body: This case study focuses on how a company can predict the uncertain future in an uncertain industry, deal with the threat of "black swan" events, and address information asymmetry. Most theories in this case study relate to these key questions.

Black swan refers to unpredictable and important events that rarely occur. They are unexpected, but they change almost everything. *Black Swan: The Impact of the Highly Improbable*, a book by Nassim Nicholas Taleb, Professor of Financial Engineering at New York University, shocked the world when it was published in 2007. According to Taleb, a black swan event has three features: it is unexpected; has extreme impacts; and can be explained or predicted even though it is unexpected. This case study clearly shows that Internet finance is a black swan for financial institutions, with the extreme impact of greatly challenging the traditional banking business. The emergence of this black swan can be explained by noting that "the traditional thinking patterns of commercial banks and financial momentum are too strong".

The information asymmetry theory, jointly analysed by three American economists, Joseph E. Stiglitz, George Akerlof and Michael Spence, holds that when various parties have different information about goods or services in economic activities, the party that has more information has the advantage, while the party with less information is at a disadvantage. Information asymmetry often leads to a substandard market place. As this case study shows, information asymmetry has existed in traditional financial markets for a long time, especially in the services that banks provided for middle, small, and micro-scale businesses. When "black swan" Internet finance challenged traditional financial institutions with disruption in payment methods, the problem of information asymmetry in the thinking patterns and operation models of traditional financial institutions got worse. At the same time, although Internet finance promoted financial innovation, it also brought financial risks, which could turn the financial markets upside down.

The development of Internet technology transformed the financial service industry, taking it in an uncertain direction. The challenge from Internet finance made the development model even more unpredictable. This means that the managers of ICBC have a lot to consider, including how to predict the uncertain future, how to deal with challenge from the "black swan", and how to address information asymmetry. This case study lists the three steps of information transformation in ICBC: adopting information transformation at every business level and implementing a transformation strategy; predicting and handling changes in Internet finance with the long-term information system independently developed by ICBC; moving from a bank that requires information transformation to a bank run on information, and reconstructing the business model with information transformation from the perspective of clients and the market. The three steps clearly show us how ICBC led a counter-attack against Internet finance by analysing strategic decision-making patterns, courses of action and ways to address the challenge of uncertainty.

Two wings: As well as addressing uncertainty with strategic insights from the Internet, this case study clearly describes the path of ICBC's information transformation, representing "one wing". 30 years of information transformation, three major upgrades of the business system, an existing fourth-generation system, the pathway for ICBC's information transformation… as well as promoting discussion about whether to develop the information system independently or outsource this task, these topics also demonstrate the theory of using information transformation to facilitate industrialization (business automation), which is an information transformation theory with Chinese characteristics. More importantly, this case study describes ICBC's efforts in recent years to address challenges that traditional information transformation theory faces. The strategic transformation of ICBC from a bank that requires information transformation to a bank run on information is not just a play on words. In fact, this transformation triggered in-depth reflection and discussion on the changes in information transformation theory. In this regard, this is an excellent case study about information transformation.

Fully integrated data, including Big Data accumulated over many years and comprehensive data from various businesses and departments, is another "wing". Viktor Mayer-Schönberger, Professor at the University of Oxford, published *Big Data: A Revolution That Will Transform How We Live, Work, and Think* in 2013. This book has been highly popular and the term "Big Data" has been frequently used ever since. However, most people only focus on how big the data is and ignore the importance of thick data and comprehensive data; they are only interested in the huge datasets created by online clicks and comments, and ignore the use of business data accumulated in IT system over many years. This case study takes ICBC as an example and explores the vital principles of Big Data application, including "Big Data analysis requires long-term planning", "Big Data analysis depends on analysing comprehensive data", "we need to improve our capacity for collecting and analysing unstructured data", and "we must eliminate information fragmentation to facilitate integration". In this case, "thick" and "comprehensive" data is applied "in the long term", which is also an excellent Big Data application.

In recent years, many people became "the boy who cried wolf" when faced with the "black swan" of Internet finance. Many case studies only briefly repeat this story, while this case study considers the digital construction of ICBC. ICBC addressed the challenge by considering the transformation of traditional thinking patterns as "one body" and IT transformation and Big Data as "two wings". With this approach, ICBC not only grasped the opportunity of the "black swan", but also "actively integrated with Internet finance" and "re-established competitiveness through digitization".

Last but not least, the end of this case study is inspiring. While most people strive for speed, Jiang Jianqing takes lessons from the ginkgo tree and the turtle, who are slow but long-lived, and recommends that managers should avoid getting over-excited and turn positive factors into stable and normal factors. Managers should also be "slow" and "stable", just like our professors at business schools.

Case Analysis II

From Internet Bank to Internet Finance

Xu Ding

ICBC is undoubtedly the largest Internet bank in the world. It has been establishing its online electronic banking system since 2002. The total number of e-banking customers has now reached 465 million, of which 193 million are online banking users and the e-banking business—mainly on Internet—makes up 87% of ICBC's overall business.

The third part of the case study (Re-establishing competitiveness through digitization) shows the strategic thinking on Internet finance by Jiang Jianqing, Chairman of ICBC, "If we can integrate funds, data and products into one data flow and play the role of information intermediaries, then IT transformation can create value. I can even see that one day the value from information transformation will be more than that from fund transformation. And that is when we become a bank that is truly run on information." With the rapid growth of Internet technologies, ICBC began its transformation from Internet banking to Internet finance.

1. Digitization of ICBC

Facing the coming flood of Internet finance, Jiang Jianqing decided to make reforms to respond to the changes and establish e-ICBC. To merge products, funds and data into one data flow, ICBC has been working against time. On 29 August, 2014, Jiang Jianqing held a morale-building meeting on Internet finance marketing in Beijing; On 23 March, 2015, he lead a financial product launch in Beijing, marking ICBC as the first bank with Internet finance brands in China, an era-defining shift for the entire Internet finance industry.

By accelerating the integration and innovation of Internet and finance, ICBC set up a complete service and operation system in less than one year, including three individual platforms for e-commerce, instant messaging and direct-sales banking business, three product lines for payment, financing and investment and an all-in-one service system that was efficient, collaborative and streamlined between online and offline. With this, e-ICBC began to take shape. Jiang once said to Masayoshi Son, President of Softbank Corporation, "ICBC will become the largest Internet corporation in the world", resolute and ambitious.

2. "Internet+" finance is making a comeback

There is one question in Internet finance which cannot be avoided: from Rakuten to Alibaba, why do e-commerce giants choose to engage in finance? First, there is a natural relationship between Internet finance and e-commerce. The e-commerce

Xu Ding, student of EMBA 2009 of CEIBS and General Manager of Private Banking Products, Department Three at the Industrial and Commercial Bank of China (ICBC).

platform and the large amounts of actual transaction data it provides offer major advantages for Internet finance; Second, as e-commerce essentially relies on finance, it is in its interest to provide more efficient and valuable finance services. In Internet strategy games between Internet companies and banks, they all come back to the same point, the essence of finance. The best proof of this is the fast rise of P2P and crowd-funding models.

With the strategic development of "Big Data and IT transformation", ICBC began to review their banking service functions. First, data mining technology has changed the local, fragmentary and decentralized information created by huge amounts of transactions in the past and makes a customer-centered bank possible. Traditional banks can only provide standardized service where customers have to follow the rules, while it is possible for customers to demand more personalized and various requests; Second, as the Internet ecosystem based on the interaction between customers, merchants and banks has taken shape, it is possible for banks to provide a one-stop and cross-cutting service. The integration of corporate and personal business, investment and fund raising, and domestic and offshore banking will become a reality. Special requests from customers may lead to a new banking service or function; Third, as the development strategy of Internet finance evolves, bold changes may take place in the business models of banks. In the future, more bank earnings may come from IT services instead of pure financial services. This means that intelligent financial services will be the development priority of banks.

3. No easy way to redefine competitiveness

In the digital era, ICBC has unprecedented opportunities and challenges to redefine its competitiveness.

First, is the existing governance structure in ICBC still suitable for the restructured business model? As ICBC used to invest heavily in branches, staffing, management tools and profit models, how will ICBC respond such operational transformation?

Second, in terms of resource allocation, ICBC can't afford the investment required to establish a social credit system, so the dream of evaluating customers through an Internet ecosystem may not be achieved for a long time.

Third, faced with the aggressive expansion of China's Internet finance, ICBC has to deal with challenges in regulation, risks and social responsibility under the spotlight of public opinion. What should ICBC do?

Fourth, concepts such as "new retail", "new corporate" and "new asset management" do nothing to address concerns about old wine in a new bottle. According to the Internet thinking, the specialized division and business process of banks will be faced with fundamental change.

Fifth, as different banks have different strategic positioning, such as universal banks or retail banks, they will focus on different areas with different Internet technologies. How should ICBC deal with destruction and reconstruction as a mega bank?

This is just the beginning of ICBC Internet finance and we look forward to seeing how the story unfolds.

References

1. Jiang J, Study on financial high-tech development and its in-depth impact, p 82
2. Wu W (2007) Cross-century strategic project—review of ICBC data centralization. China Urban Finance, vol 3
3. Yu T (2013) Ali, tencent and JD make the internet popular for one year. PE Daily, 21 Mar 2013. http://chuangye.cyz.org.cn/2013/0321/36263.shtml
4. Ma W (2014) The internet finance cannot subvert commercial banks! ForbesChina, 24 May 2014. http://www.forbeschina.com/review/201405/0033248_7.shtml

Epilogue

Iteration of Cases Through Time

Throughout the process of compiling this book, we have focused relentlessly on new business models and new science and technology. However, as time goes by, some new case studies may become outdated.

Alibaba: The Decade-long Road to Financial Services was written a year ago and well received by students. When preparing for this year's classes, I felt that it was necessary to make some modifications. We collected about 50 news items (see Exhibit 1) from yicai.com and the official website of Ant Financial for the outline of *Ant Financial: One Year Progress in Financial Services*, in an effort to make it a sequel to *Alibaba: The Decade-Long Road to Financial Services*[1].

On 24 October, 2015, during an EMBA class, I invited two students to share their versions of the case study outline completed in advance, and I also made a copy of mine and added it to the teaching materials (see Exhibit 2). This brought me an exciting and fresh new experience of case study iteration.

After that, I continued to share with students the core content of "Ant Financial Cloud" in 10 PPT slides (consisting of pictures, graphs and videos), which include: (1) the "Ant Financial Cloud" provides SMEs with agile cloud computing services through Distributed Transaction Services, Distributed Data Source Services, Distributed Messaging Services, Distributed Scheduling Services and the Distributed Service Registry; (2) Inspired by 3D printing technology that allows easy printing of customized objects, Ant Financial has been exploring the ways to leverage "component transformation" to produce customized financial apps for SMEs. This made us think: With today's booming Internet finance, why did Ant

---

[1]*Alibaba: The Decade-long Road to Financial Services* was authorized by the Alibaba Group to be used as a teaching case study. While the composition of *Ant Financial: One Year Progress of Financial Services* was completed based on 51 pieces of information published on the official website of Ant Financial and yicai.com within a period of one year (2014/10/16–2015/10/6) starting from its foundation. It is not entirely based on specific research and has not been approved by Ant Financial.

© Springer Nature Singapore Pte Ltd and Shanghai Jiao Tong University Press 2018   173
X. Zhu, *China's Technology Innovators*, Management for Professionals,
DOI 10.1007/978-981-10-5388-7

Financial treat the "cloud" as its core capacity for differentiated competition with both traditional financial institutions and Internet-based financial institutions? The discussion also brought us a deeper, real and more up-to-date experience of case study iteration.

On 28 October, 4 days after that class, chinanews.com reported that AliCloud had managed to sort 100TB of data in 377 seconds, setting a new world record in the Sort Benchmark contests, the world computing "Olympics". The fast iteration of technological competition brings both challenges and opportunities for the iteration of case studies in teaching.

In the epilogue of *The Big Switch*, Nicholas Carr wrote that, "All technological change is generational change. The full power and consequences of a new technology are unleashed only when those who have grown up with it become adults and begin to push their outdated parents to the margins... It's in this way that progress covers its tracks, perpetually refreshing the illusion that where we are is where we were meant to be..."

However, the case study iteration of CEIBS may need to be counted on a yearly basis, rather than by "generation", because events and trends change at a far faster speed than we ever could imagine.

Finally, we should specially thank Li Mingjun (李铭俊), Pedro Nueno (佩德罗•雷诺), Ding Yuan (丁　远), Zhang Weijiong (张维炯), Xu Dingbo (许定波), Zhou Xuelin (周雪林); Liang Neng (梁　能), Xu Leiping (许雷平), Hu Zhifeng (胡峙峰), Song Yanbo (宋彦博), Ni Yingzi (倪英子), Zhu Qiong (朱　琼), Ren Yifan (任轶凡), Ji Chendong (季宸东), Li Yang (李　杨), Liu Fei (柳　飞); Fan Xiaojun (范小军), Ye Weiling (叶巍岭), Dong Ming (董　明), Xu Shujun (许淑君), Li Yugang (李玉刚), Yu Xiubao (俞秀宝), Lao Guoling (劳帼龄), Wang Liping (王理平), He jiaxun (何佳讯); Lian Minling (连敏玲), Sun Zikui (孙子奎), Xu Qiang (徐　强), Li Yuan (李源), Zhang Bofan (张博凡), Zhao Zhigang (赵志刚), Xu Ding (徐　鼎), Li Tianjun (李天军), Shen Jianfang (沈建芳); Chen Jieping (陈杰平), Su Xijia (苏锡嘉), Zhao Xiaolei (赵筱蕾), Lai Weidong (赖卫东), Mao Zhuchen (毛竹晨), Wang Qingjiang (王庆江), Fu Danyang (傅丹阳), Zhang Lin (张　琳); Xiong Lei (熊　磊), Li Yishi (李逸石), Yang Xiaoping (杨小平), Wang Qi (王　琦), Li Zhenhua (李振华), Zhou Pin (周　频), Tang Wen (唐　雯); Cao Xuehui (曹雪会), Huang Chengyan (黄成彦), Zhu Yezi (朱叶子), Shi Tianyu (施天瑜), Xiao Yingjun (肖颖君), Xu Jianmin (徐建敏), Jiang Junzhe (姜浚哲), Zhang Yu (张　羽), Fan Jingjing (范晶晶), Ma Lan (马蓝), Song Bingying (宋炳颖), Wang Chengde (汪承德), Wang Danping (王丹萍), Li Rui (李　蕊), Guan Peng (关　鹏), Zhu Yifan (朱奕帆), Zhang Yingwen (张颖文).

Dr. Xiaoming Zhu, Professor of Management at CEIBS

Epilogue—Exhibit 1

Collection of Alibaba and Ant Financial News from 2014–2015[2]

I. 2014

1. 16 October	Establishment of Ant Financial
2. 27 October	The number of users of Yu'e Bao increased to 149 million
3. 20 November	Guangzhou initiated a pilot program using Alipay for medical payment
4. 2 December	Alipay launched overseas transport card service
5. 9 December	China Insurance Regulatory Commission (CIRC) approved Ant Financial's asset support plan for small loans, which was the first of its kind
6. 10 December	Alipay obtained Apple's interface for mobile fingerprint payment
7. 19 December	Alipay partnered with China Smartpay (8325.HK) to promote Alipay Wallet's face-to-face payment service in Asia

II. 2015

8. 26 January	Alipay added the red envelope function
9. 26 January	Zhima Credit passed a public beta test, entering the individual credit reference market
10. 10 February	Alibaba officially incorporated its small loan business into Ant Financial
11. 12 February	Ant Financial purchased a stake in the Tebon Fund
12. 26 February	South Korea's Financial Supervisory Service (FSS) approved a cooperation between Alipay and Hana Bank to jointly launch a payment service for Chinese tourists in South Korea
13. 31 March	Alipay launched "City Services" (such as marriage registration, payment, and exit and entry inquiries)
14. 8 April	Zhima Credit entered the consumer finance market for the first time
15. 8 April	Alibaba Pictures integrated Yulebao with Taobao Movie in a move to build an online entertainment platform
16. 9 April	Ant Financial released the "CSI Taojin 100" with the China Securities Index (CSI) which created a new trend of big-data-based stock selection

---

[2]*Source* yicai.com and reports from the Ant Financial official website.

17. 12 April   AliCloud opened a data centre in Silicon Valley, USA, marking the start of the global expansion of Chinese cloud computing companies

18. 13 April   "Greenland Dichan Bao", the first Internet-based financial product of the real estate industry, was jointly launched by Ant Financial, Greenland Group and lu.com of Ping An

19. 24 April   Ant Financial invested 200 million in a controlling stake of Fund 123 (an online mutual fund) to boost the fund business

20. 5 May   The cumulative turnover of Zhao Caibao exceeded 100 billion

21. 11 May   Zhao Caibao partnered with China Orient Asset Management Co., Ltd. (COMAC) to work on comprehensive financial services for investment, financing and other business areas such as credit enhancement, credit rating, and investment platforms

22. 19 May   Ant Financial launched the equity crowd-funding platform —"Antsdaq"

23. 20 May   Zhima Credit was introduced into baihe.com, taking blind dating into the "credit dating" era

24. 27 May   Part of Alipay's optical fibre was cut off in Xiaoshan, Hangzhou

25. 2 June   Zhima Credit reached strategic data cooperation with Rong 360, in order to participate in the online credit reference market

26. 4 June   Users of Zhima Credit were allowed to apply for visas in Singapore and Luxembourg based on their scores

27. 8 June   Ant Financial obtained a 20.62% stake in Hundsun Electronics through its acquisition of Zhejiang Rongxin Technology Development Co., Ltd

28. 8 June   Ant Financial joined hands with yicai.com to establish the Internet Finance Institute

29. 16 June   Ant Financial and the Shanghai Airport Authority entered into a strategic framework agreement to allow one-stop boarding with Alipay Wallet

30. 18 June   The National Council for Social Security Funda (NSSF) invested in Ant Financial and obtained a 5% of stake

31. 23 June   Koubei.com was founded with 6 billion of joint investment from Alibaba and Ant Financial

32. 25 June   Mybank, an online private bank, was launched

33. 30 June   Weshare Finance announced that Shandian Jiekuan had reached a strategic cooperation agreement with Zhima Credit to accelerate expansion in the individual credit reference market

34. 30 June   The number of merchants supporting Alipay hit 130,000

35. 1 July	Alipay partnered with KFC to enable payment via the Alipay app
36. 2 July	Ant Small Loans offered the first purely credit-based loan to farmers
37. 3 July	Ant Financial closed its first round of funding with Jack Ma holding about 7% of the stake
38. 8 July	Alipay released version 9.0, adding new functions of "Merchant" and "Friends"
39. 24 July	"Ant Credit" of Ant Financial cooperated with over 40 online shopping platforms such as Amazon.com and dianping.com to support payment by "Ant Credit"
40. 31 July	Alipay Wallet was about to add a "Stock" function at the end of July, allowing users to directly buy stocks with money from Yu'e Bao
41. 7 August	The China Southern Power Grid joined hands with Ant Financial and AliCloud to provide a city power grid service under the "Internet+" strategy
42. 11 August	Ant Financial led investment in "Qudian" (formerly known as Qufenqi), an online shopping centre for installment purchases by college students
43. 27 August	eastday.com signed a contract with Ant Financial to jointly develop urban livelihood services
44. 31 August	Ant Financial launched an independent wealth management platform—Ant Fortune
45. 10 September	The China Post purchased a stake of Ant Financial
46. 16 September	Passengers with a score of 750 or more in Zhima Credit were allowed to exit through the VIP channel at Beijing Capital International Airport
47. 18 September	Ant Financial secured a controlling stake of Cathay Insurance with 1.2 billion
48. 14 October	Ant Financial (Alipay) officially deployed real-time data protection
49. 15 October	Ant Financial made a strategic move to buy a stake in 36Kr
50. 16 October	Ant Financial implemented the project of transparent "provisions" regulation
51. 16 October	Ant Financial announced the "Internet Booster Plan" and opened the Ant Financial Cloud to the outside world

Epilogue—Exhibit 2

Ant Financial: One Year Progress of Financial Services (outline)

I. Introduction

There is a famous quote by the poet Jia Dao of the Tang Dynasty (618-907 AD) that "It takes a decade to mould a fine sword". The case study—*Alibaba: The Decade-long Road to Financial Services* (CEIBS 2014) includes the "Seven Swords Strategy" created by Alibaba over the period of a decade. However, the story of Alibaba continued. In October 2014, "Ant Financial" was established. This news was reported by numerous media throughout that year. In this time, yicai.com and the official Alibaba website published 51 pieces of related news, which broadly depict the overall "progress in an entire year" of Ant Financial.

II. The Hard-Working Ant Financial

1. Livelihood services: Alipay engaged in medical insurance payment, providing services of marriage registration, payment, exit and entry inquiry, as well as applications for Singapore and Luxembourg visas. Alipay Wallet also began providing one-stop boarding at Shanghai airports; and Alipay established cooperation with eastday.com.
2. International services: Alipay began offering overseas transport cards, partnered with Smartpay to promote face-to-face payment in Asia; it was also approved to provide payment services for Chinese tourists in South Korea.
3. Service expansion: Alipay added the Red Envelope function; Yulebao and Taobao Movie built an online entertainment platform; the cumulative turnover of ZhaoCaibao reached 350 billion; Ant Financial launched the Ant Fortune.
4. Service cooperation: Ant Financial bought a stake of Tianhong Asset Management, the Tebon Fund and Cathay Insurance; it partnered with private companies and established Mybank.
5. Service innovation: Ant Financial launched "CSI Taojin 100", forming a new trend; Alipay obtained Apple's interface for mobile fingerprint payment to enhance the connection of financial services with other industries and for the purpose of risk management.

Cainiao (rookie), zhima (sesame) and ant are quite ordinary concepts, but they have some valuable qualities. For example, ants are hard-working and always work as a team, helping work to get done more easily. This is well reflected in the five kinds of services offered by Ant Financial.

III. The Fast-Developing Ant Financial

1. Becoming an innovation centre: the Ant Financial Cloud is currently the most mature application for cloud computing in finance. It supported the construction of the first fully cloud-based bank Mybank, and the scenario-based insurance innovation of Zhong An Insurance, which owns half of the total policies of the insurance industry and relies on the support of the financial cloud; Alipay managed to resume its business shortly after the optical fibre broke down, showing the technological innovation capability of Ant Financial.
2. Going global: AliCloud was the first cloud service centre built by a Chinese company in Silicon Valley. Alipay+ is promoting its services to the world, including the USA, Australia, India, etc.
3. Moving towards win-win cooperation: forming partnerships with over 200 financial institutions; joining hand with caijing.com and establishing the Internet Finance Institute; working with the China Southern Power Grid, providing an urban power grid service.
4. Moving towards inclusive finance: the average amount of savings of Yu'e Bao's 180 million users is about RMB 4000, and Ant Fortune allows a minimum of RMB 100 investment, bringing users the joy of wealth management; Alibaba incorporated its small loan business into Ant Financial and shelled out 600 million to establish koubei.com; Ant Small Loans offered purely credit-based loans to farmers for the first time.
5. Moving forward with rules and compliance: Zhima Credit entered into consumer finance for the first time; it strictly followed the rules and regulations of "one bank and three commissions" (People's Bank of China, China Securities Regulatory Commission, China Banking Regulatory Commission and China Insurance Regulatory Commission) in conducting business, for example, it obtained the license for consigned funds before acquiring Fund 123.

These five directions show the resolution of Ant Financial in carrying out financial innovation. As we all know, credit is the foundation of financial service regulations. Along with the release of the Non-Bank Payment Institutions' Online Payment Service Management Approach (Draft for Comments), both Ant Financial and the "booming" Internet finance industry will face the test of regulation and compliance.

IV. Conclusion

Technology is neutral, but people have good and bad intentions, so when Internet promotes the "good intentions" of a contribution, it also promotes certain "bad intentions". There is now an urgent need to strike a balance between the development and governance of financial innovation. In the end, the financial services industry will become orderly, and shift from being wealth-oriented to inclusive finance, and from being profit-driven to service-driven. Born out of Alibaba, Ant Financial has accomplished many achievements in just one year. It is now moving fast and steadily forward into the future. In the future, amid challenges and opportunities, no doubt Ant Financial will have highs and lows. So it should maintain its financial service company attributes and respond to strong competition by pursuing the reliability and consistency of financial software. You will only be successful at the end of an amazing journey if you focus on your original goal.

Dr. Zhu Xiaoming, Professor of Management at CEIBS